D0007455

THE
MATHEMATICAL
MECHANIC

MARK LEVI

THE
MATHEMATICAL
MECHANIC

USING
PHYSICAL
REASONING
TO SOLVE
PROBLEMS

PRINCETON UNIVERSITY PRESS   PRINCETON AND OXFORD

Copyright © 2009 by Princeton University Press

Published by Princeton University Press, 41 William Street,
Princeton, New Jersey 08540

In the United Kingdom: Princeton University Press, 6 Oxford Street,
Woodstock, Oxfordshire OX20 1TW

All Rights Reserved

Library of Congress Cataloging-in-Publication Data
Levi, Mark, 1951–
The mathematical mechanic: using physical reasoning to
solve problems / Mark Levi.
p.   cm.
Includes bibliographical references and index.
ISBN 978-0-691-14020-9 (cloth : alk. paper)
1. Problem solving.   2. Mathematical physics.   I. Title.
QA63.L48 2009
510–dc22   2009004861

British Library Cataloging-in-Publication Data is available

This book has been composed in Times

Printed on acid-free paper. ∞

press.princeton.edu

Typeset by S R Nova Pvt Ltd, Bangalore, India

Printed in the United States of America

10   9   8   7   6   5   4   3   2   1

Contents

THE
MATHEMATICAL
MECHANIC

1

INTRODUCTION

IT SO HAPPENS THAT ONE OF
THE GREATEST MATHEMATICAL
DISCOVERIES OF ALL TIMES
WAS GUIDED BY PHYSICAL
INTUITION.
—GEORGE POLYA, ON
ARCHIMEDES' DISCOVERY OF
INTEGRAL CALCULUS

1.1 Math versus Physics

Back in the Soviet Union in the early 1970s, our undergraduate class—about forty mathematics and physics sophomores—was drafted for a summer job in the countryside. Our job included mixing concrete and constructing silos on one of the collective farms. My friend Anatole and I were detailed to shovel gravel. The finals were just behind us and we felt free (as free as one could feel in the circumstances). Anatole's major was physics; mine was mathematics. Like the fans of two rival teams, each of us tried to convince the other that his field was superior. Anatole said bluntly that mathematics is a servant of physics. I countered that mathematics can exist without physics and not the other way around. Theorems, I added, are permanent. Physical theories come and go. Although I did not volunteer this information to Anatole, my own reason for majoring in mathematics was to learn the main *tool* of physics—the field which I had planned to eventually pursue. In fact, the summer between high school and college I had bumped into my high school physics teacher, who asked me about my plans for the Fall. "Starting on my math major," I said. "What? Mathematics? You are nuts!" was his reply. I took it as a compliment (perhaps proving his point).

1.2 What This Book Is About

This is not "one of those big, fat paperbacks, intended to while away a monsoon or two, which, if thrown with a good overarm action, will bring a water buffalo to its knees" (Nancy Banks-Smith, a British television critic). With its small weight this book will not bring people to their knees, at least not by its *physical* impact. However, the book does exact revenge—or maybe just administers a pinprick—against the view that mathematics is a servant of physics. In this book physics is put to work for mathematics, proving to be a very efficient servant (with apologies to physicists). Physical ideas can be real eye-openers and can suggest a strikingly simplified solution to a mathematical problem. The two subjects are so intimately intertwined that both suffer if separated. An occasional role reversal can be very fruitful, as this book illustrates. It may be argued that the separation of the two subjects is artificial.[1]

Some history. The Physical approach to mathematics goes back at least to Archimedes (c. 287 BC – c. 212 BC), who proved his famous integral calculus theorem on the volumes of the cylinder, a sphere, and a cone using an imagined balancing scale. The sketch of this theorem was engraved on his tombstone. Archimedes' approach can be found in [P]. For Newton, the two subjects were one. The books [U] and [BB] present very nice physical solutions of mathematical problems. Many of fundamental mathematical discoveries (Hamilton, Riemann, Lagrange, Jacobi, Möbius, Grassmann, Poincaré) were guided by physical considerations.

Is there a general recipe to the physical approach? As with any tool—physical[2] or intellectual—this one sometimes works and sometimes does not. The main difficulty is to come up with a

---

[1]"Mathematics is the branch of theoretical physics where the experiments are cheap" (V. Arnold [ARN]). Not only are the experiments in this book cheap—they are even free, being the thought experiments (see, for instance, problems 2.2, 3.3, 3.13, and, in fact, most of the problems in this book).

[2]With apologies for the pun.

physical incarnation of the problem.[3] Some problems are well suited for this treatment, and some are not (naturally, this book includes only the former kind). Finding a physical interpretation of a particular problem is sometimes easy, and sometimes not; readers can form their own opinions by skimming through these pages.

One lesson a student can take from this book is that looking for a physical meaning in mathematics can pay off.

Mathematical rigor. Our physical arguments are not rigorous, as they stand. Rather, these arguments are sketches of rigorous proofs, expressed in physical terms. I translated these physical "proofs" into mathematical proofs only for a few selected problems. Doing so systematically would have turned this book into a "big, fat ...". I hope that the reader will see the pattern and, if interested, will be able to treat the cases I did not treat. Having made this disclaimer I feel less guilty about using the word "proof" throughout the text without quotation marks.

The main point here is that the physical argument can be a tool of discovery and of intuitive insight—the two steps preceding rigor. As Archimedes wrote, "For of course it is easier to establish a proof if one has in this way previously obtained a conception of the question, than for him to seek it without such a preliminary notion" ([ARC], p. 8).

An axiomatic approach. Instead of translating each physical "proof" into a rigorous proof, an interesting project would entail systematically developing "physical axioms"—a set of axioms equivalent to Euclidean geometry/calculus—and then repeating the proofs given here in the new setting.

One can imagine an extraterrestrial civilization that first developed mechanics as a rigorous and pure axiomatic subject. In this dual world, someone would have written a book on using geometry to prove mechanical theorems.

Perhaps the real lesson is that one should not focus solely on one or the other approach, but rather look at both sides of the coin. This

---

[3]It is a contrarian approach: normally one starts with a physical problem, and abstracts it to a mathematical one; here we go in the opposite direction.

book is a reaction to the prevalent neglect of the physical aspect of mathematics.

Some psychology. Physical solutions from this book can be translated into mathematical language. However, something would be lost in this translation. Mechanical intuition is a basic attribute of our intellect, as basic as our geometrical imagination, and not to use it is to neglect a powerful tool we possess. Mechanics is geometry with the emphasis on motion and touch. In the latter two respects, mechanics gives us an extra dimension of perception. It is this that allows us to view mathematics from a different angle, as described in this book.

There is a sad Darwinian principle at work. Physical reasoning was responsible for some fundamental mathematical discoveries, from Archimedes, to Riemann, to Poincaré, and up to the present day. As a subject develops, however, this heuristic reasoning becomes forgotten. As a result, students are often unaware of the intuitive foundations of subjects they study.

The intended audience. If you are interested in mathematics and physics you will, I hope, not toss this book away.

This book may interest anyone who thinks it is fascinating that

- The Pythagorean theorem can be explained by the law of conservation of energy.
- Flipping a switch in a simple circuit proves the inequality $\sqrt{ab} \leq \frac{1}{2}(a+b)$.
- Some difficult calculus problems can be solved easily with no calculus.
- Examining the motion of a bike wheel proves the Gauss-Bonnet formula (no prior exposure is assumed; all the background is provided).
- Both the Riemann integral formula and the Riemann mapping theorem (both explained in the appropriate section) become intuitively obvious by observing fluid motion.

This book should appeal to anyone curious about geometry or mechanics, or to many people who are not interested in mathematics because they find it dry or boring.

Uses in courses. Besides its entertainment value, this book can be used as a supplement in courses in calculus, geometry, and teacher education. Professors of mathematics and physics may find some problems and observations to be useful in their teaching.

Required background. Most of the book (chapters 2–5) requires only precalculus and some basic geometry, and the level of difficulty stays roughly flat throughout those chapters, with a few crests and valleys. Chapters 6 and 7 require only an acquaintance with the derivative and the integral. At the end of chapter 7 I mention the divergence, but in a way that requires no prior exposure. This chapter should be accessible to anyone familiar with precalculus.

The second part (chapters 6–11) uses on rare occasions a few concepts from multivariable calculus, but I tried to avoid the jargon as much as possible, hoping that intuition will help the reader jump over some technical gaps.

Everything one needs from physics is described in the appendix; no prior background is assumed.

This book can be read one section or problem at a time; if you get stuck, it only takes turning a page to gain traction. A few exceptions to this topic-per-page structure occur, mostly in the later chapters.

Sources. Many, but but not all *solutions* in this book are, to my knowledge, original. These include solutions to problems 2.6, 2.9, 2.10, 2.11, 2.13, 3.3, 3.7, 3.8, 3.9, 3.10, 3.11, 3.12, 3.17, 3.18, 3.19, 3.20, 3.21, 5.2, 5.3, 6.1, 6.2, 6.3, 6.4, 6.5, 7.1, and 7.2. The interpretations in chapter 8 and in sections 9.3, 9.8 and 11.8 appear to be new.

There is not much literature on the topic of this book. When I was in high school, an example from Uspenski's book [U] struck me so much that the topic became a hobby.[4] More problems of the

---

[4]This is the first example of this book, in section 2.2. Tokieda's article [TO] contains, together with this example, some very nice additional ones.

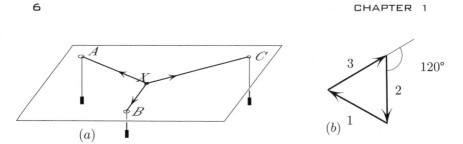

Figure 1.1. If X minimizes total distance $XA + XB + XC$, then the angles at X are 120°.

kind described here are in the small book by Kogan [K] and Balk and Boltyanskii [BB], and in chapter 9 of Polya's book [P]. And the main source of such problems and solutions is the 24-centuries-old work by Archimedes [ARC].

1.3 A Physical versus a Mathematical Solution: An Example

Problem. *Given three points A, B, and C in the plane, find the point X for which the sum of distances $XA + XB + XC$ is minimal.*

Physical approach. We start by drilling three holes at A, B, and C in a tabletop (this is cheaper to do as a thought experiment or at a friend's home). Having tied the three strings together, calling the common point X, I slip each string through a different hole and hang equal weights under the table, as shown in figure 1.1. Let us make each weight equal to 1; the potential energy of the first string is then AX: indeed, to drag X from the hole A to its current position X we have to raise the unit weight by distance AX. We endowed the sum of distances $XA + XB + XC$ with the physical meaning of potential energy. Now, if this length/energy is minimal, then the system is in equilibrium. The three forces of tension acting on X then add up to zero and hence they form a triangle (rather than an open path) if placed head-to-tail, as shown in figure 1.1(b). This triangle is equilateral since the weights are equal, and hence the angle between positive directions of these vectors is 120°. We showed that $\angle AXB = \angle BXC = \angle CXA = 120°$.

Mathematical solution. Let **a**, **b**, **c**, and **x** denote the position vectors of the points A, B, C, and X respectively. We have to minimize the sum of lengths $S(\mathbf{x}) = |\mathbf{x} - \mathbf{a}| + |\mathbf{x} - \mathbf{a}| + |\mathbf{x} - \mathbf{a}|$. To that end, we set partial derivatives of S to zero: $\frac{\partial S}{\partial x} = \frac{\partial S}{\partial y} = 0$, where $\mathbf{x} = (x, y)$, or, expressing the same condition more compactly and geometrically, we set the gradient $\nabla S = \langle \frac{\partial S}{\partial x}, \frac{\partial S}{\partial y} \rangle = \mathbf{0}$. We now compute ∇S. We have $\frac{\partial}{\partial x}|\mathbf{x} - \mathbf{a}| = \frac{\partial}{\partial x}\sqrt{(x - a_1)^2 + (y - a_2)^2} = (x - a_1)/\sqrt{(x - a_1)^2 + (y - a_2)^2}$, and similarly $\frac{\partial}{\partial y}|\mathbf{x} - \mathbf{a}| = (y - a_2)/\sqrt{(x - a_1)^2 + (y - a_2)^2}$. Thus $\nabla|\mathbf{x}-\mathbf{a}| = (\mathbf{x}-\mathbf{a})/|\mathbf{x}-\mathbf{a}|$ is a unit vector, pointing from A to X. We will denote this vector by $\mathbf{e}_a$. This result came from an explicit calculation, but its physical meaning, borrowed from the physical approach, is simply the force with which X pulls the string. Differentiating the remaining two terms $|\mathbf{x}-\mathbf{b}|$ and $|\mathbf{x}-\mathbf{c}|$ in S we obtain $\nabla S = \mathbf{e}_a + \mathbf{e}_b + \mathbf{e}_c$, where $\mathbf{e}_b$ and $\mathbf{e}_c$ are defined similarly to $\mathbf{e}_a$. We conclude that the optimal position X corresponds to $\nabla S = \mathbf{e}_a + \mathbf{e}_b + \mathbf{e}_c = \mathbf{0}$. Thus the unit vectors $\mathbf{e}_a$, $\mathbf{e}_b$, $\mathbf{e}_c$ form an equilateral triangle, and any exterior angle of that triangle, that is, the angle between any pair of our unit vectors, is 120°.

It is fascinating to observe how the difficulty changes shape in passing from one approach to the other. In the mathematical solution, the work goes into a formal manipulation. In the physical approach, the work goes into inventing the right physical model. This pattern is shared by many problems in this book.

Relative advantages of the two approaches.

Physical approach	Mathematical approach
Less or no computation	Universal applicability
Answer is often conceptual	Rigor
Can lead to new discoveries	
Less background is required	
Accessible to precalc students	

The physical approach suits some subjects more than others. The subject of complex variables is one example where physical intuition is very fruitful. Some of the fundamental ideas of the subject, such as the Cauchy-Goursat theorem, the Cauchy integral formula, and the Riemann mapping theorem, can be made intuitively obvious in a short time, with minimal physical background. With these ideas Euler's formula

$$1 + \frac{1}{4} + \frac{1}{9} + \cdots + \frac{1}{n^2} + \cdots = \frac{\pi^2}{6}$$

acquires a nice interpretation, saying that, for a special incompressible fluid flow in the plane, the fluid injected at the origin at the rate of $\frac{\pi^2}{6}$ gallons per second is absorbed entirely by sinks located at integer points (the details are given in section 11.8 on complex variables). Many such examples can be found in other fields of mathematics, and I hope more will be written on this in the future.

1.4 Acknowledgments

This book would probably not have been written had it not been for something my father said when I was 16. I showed him a physical paradox that had occurred to me, and he said: "Why don't you write it down and start a collection?" This book is an excerpt from this collection, with a few additions.

Many of my friends and colleagues contributed to this book by suggestions and advice. I thank in particular Andrew Belmonte, Alain Chenciner, Charles Conley, Phil Holmes, Nancy Kopell, Paul Nahin, Sergei Tabachnikov, and Tadashi Tokieda. Thanks to their stimulation the collection was massaged into a presentable form. I am in particular debt to Andy Ruina, who read much of the manuscript and made many suggestions and corrections. I am grateful to Anna Pierrehumbert for her numerous suggestions which improved this book, and to Vickie Kearn for her encouragement.

I gratefully acknowledge support by the National Science Foundation under Grant No. 0605878.

2

THE PYTHAGOREAN THEOREM

2.1 Introduction

Here is a fact seemingly not worth mentioning for its triviality: **Still water in a resting container, with no disturbances, shall remain at rest.** I think it is remarkable that this fact has the Pythagorean theorem as a corollary (p. 17). In addition, this seeming triviality implies the law of sines (p. 18), the Archimedian buoyancy law, and the 3D area version of the Pythagorean theorem (p. 19).

The proof of the Pythagorean theorem, described in section 2.2, suggested a kinematic proof of the Pythagorean theorem, described in section 2.6. The motion-based approach makes some other topics very transparent, including

- The fundamental theorem of calculus.
- The computational formula for the determinant.
- The expansion of the determinant in a row.

All these are described in this chapter.

Several more physical proofs of the Pythagorean theorem are given here, one using springs, and the other using kinetic energy.

The unifying theme of this chapter is the Pythagorean theorem, although we do go off on a few short tangents.

2.2 The "Fish Tank" Proof of the Pythagorean Theorem

Let us build a prism-shaped "fish tank" with our right triangle as the base (figure 2.1). We mount the tank so that it can rotate freely

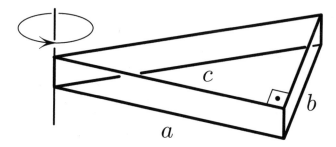

Figure 2.1. The water-filled fish tank, free to rotate around a vertical edge, has no desire to.

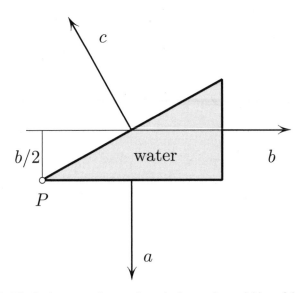

Figure 2.2. The Pythagorean theorem is equivalent to the vanishing of the combined torque upon the tank around P.

around the vertical axis through one end of the hypotenuse. Now let us fill our fish tank with water.

The water pushes on the walls in three competing directions as figure 2.2 shows, each force trying to rotate the tank around P. Of course, the competition is a draw: the tank has zero desire to rotate. Otherwise we would have had an engine which uses no fuel—a so-called perpetual motion machine, forbidden by the law of conservation of energy.

In this case the "desire" is the sum of the three torques of the pressure forces. We note here[1] that the torque of the force around a pivot point P is simply the force's magnitude times the distance from the line of force to the pivot point. The torque measures the intensity with which the force tries to rotate the object it's applied to around P.

For convenience, let us assume the force of pressure to be 1 pound per unit length of the wall—we can always achieve it by adjusting water depth. The three forces are then a, b, and c; the corresponding levers are $a/2$, $b/2$, and $c/2$, and the zero torque condition reads

$$a \cdot \frac{a}{2} + b \cdot \frac{b}{2} - c \cdot \frac{c}{2} = 0, \tag{2.1}$$

or $a^2 + b^2 = c^2$, giving us the Pythagorean theorem!

Still water. Note that we didn't have to build the fish tank, not even in the thought experiment; rather, we can imagine the prism of water embedded in a larger body of water. The Pythagorean theorem follows as before from the fact that the prism will not spontaneously rotate under the pressure of the surrounding fluid on its vertical faces. We conclude that the Pythagorean theorem is a consequence of the fact that still water remains still.

Exercise. *From a point A outside a circle draw a tangent line AT and a secant line APQ as shown in figure 2.3. Prove that*

$$AP \cdot AQ = AT^2. \tag{2.2}$$

Hint: Consider the shaded curvilinear triangle APT in figure 2.3, thought of as a rigid container filled with gas and allowed to pivot around O.

As explained in section 2.3 in a different context, (2.2) expresses the fact that the shaded area remains unchanged under rotations around O. Similarly, the Pythagorean theorem expresses the fact that the area of a right triangle remains unchanged as the triangle is rotated around one of the ends of the hypotenuse.

---

[1] See section A.5 for full background.

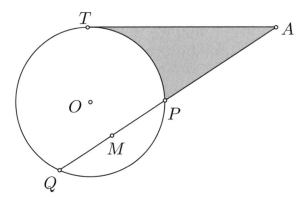

Figure 2.3. Proving $AP \cdot AQ = AT^2$.

2.3 Converting a Physical Argument into a Rigorous Proof

The pivotal[2] point of the "fish tank" proof of the Pythagorean theorem was the vanishing of the net torque around P (figure 2.1). How can we restate this zero-torque idea in purely mathematical terms, without appealing to physical concepts? Here is the answer.

The *physical* statement (2.1) of zero net torque around P translates into the *geometrical* statement that the area of the triangle does not change when the triangle is rotated around P.[3] Here is the proof of this equivalence.

Let $A(\theta)$ be the area of the triangle rotated around P through the angle θ. This area is, of course, independent of θ:

$$A'(\theta) = 0,$$

and we claim that it is this constancy of the area that is equivalent to the zero-torque condition (2.1). To show this equivalence it suffices to show that

$$A'(\theta) = a \cdot \frac{a}{2} + b \cdot \frac{b}{2} - c \cdot \frac{c}{2}. \qquad (2.3)$$

[2]This pun was not originally intended.
[3]Here is an example where a trivial-sounding fact (the area of the triangle doesn't change under rotations) hides something less trivial (the Pythagorean theorem.)

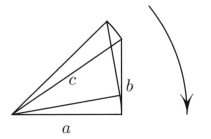

Figure 2.4. The area swept by the two legs equals the area swept by the hypotenuse.

To demonstrate (2.3) we rotate the triangle through a small angle $\Delta\theta$ around P. The side a sweeps a sector of area $\frac{1}{2}a^2\Delta\theta$, with a similar expression for c. In fact, the area swept by b is given by the same expression: $\frac{1}{2}b^2\Delta\theta$. Indeed, b executes two motions simultaneously: (i) sliding in its own direction, contributing nothing to the rate of sweeping of the area, and (ii) rotation around its leading end. We conclude that the area swept is $\frac{1}{2}b^2\Delta\theta$. The total area swept by all three sides is

$$\Delta A = \left(\frac{1}{2}a^2 + \frac{1}{2}b^2 - \frac{1}{2}c^2 \right) \Delta\theta.$$

Here the minus sign is due to the fact that the area is "lost" through the hypotenuse. Dividing by $\Delta\theta$ and taking the limit as $\Delta\theta \to 0$, we obtain (2.3).

Here are a few other applications of the idea of sweeping:

1. A "ring" proof of the Pythagorean theorem described in section 2.6.
2. A remark on the area between the tracks of two wheels of a bike (section 6.1).
3. A visual proof that the determinant $\left| \begin{smallmatrix} a & b \\ c & d \end{smallmatrix} \right| = ad - bc$ equals the area of a parallelogram generated by the vectors $\langle a, c \rangle$ and $\langle b, d \rangle$ (section 2.5).
4. A visual proof of the formula for the row decomposition of a determinant (section 2.5).

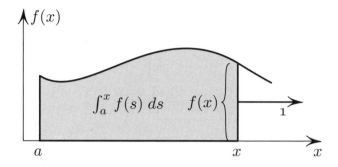

Figure 2.5. The fundamental theorem of calculus: the area changes at the rate equal
to the length $f(x)$ of the moving boundary times its speed (1).

2.4 The Fundamental Theorem of Calculus

The idea of considering the areas swept by a moving segment is very
fruitful. In fact, the fundamental theorem of calculus

$$\frac{d}{dx} \int_a^x f(s)\, ds = f(x)$$

is an example; the theorem says that *a segment moving with unit
speed in the direction perpendicular to itself sweeps the area at the
rate equal to the segment's length ($f(x)$) times its speed* (1).

The same idea applies to the integral with both ends variable, even
with compound dependence. For example, we can immediately see
that

$$\frac{d}{dt} \int_{g(t)}^b f(s)\, ds = -f(g(t))\, g'(t)$$

by repeating the preceding italicized sentence: the rate of change of
the area equals the product of the length $f(g(t))$ of the moving front
and its velocity $-g'(t)$. The minus sign is due to the fact that the
boundary moves inward: the positive direction moves outward.

We could allow the upper end to depend on time as well, leaving
the justification as an exercise:

$$\frac{d}{dt} \int_{g(t)}^{h(t)} f(s)\, ds = f(h(t))\, h'(t) - f(g(t))\, g'(t).$$

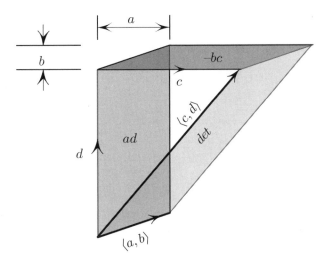

Figure 2.6. The segment moves while remaining parallel to itself. The area swept does not depend on the path of the segment.

2.5 The Determinant by Sweeping

The determinant $\left|\begin{smallmatrix} a & b \\ c & d \end{smallmatrix}\right|$ is, by defnition, the area of the parallelogram generated by the vectors $\langle a, b \rangle$ and $\langle c, d \rangle$. This definition leads to the computational formula,[4] giving the value $ad - bc$. Here is a kinematic explanation of this formula, due to Nana Wang, using again the fruitful idea of sweeping.

The area in question is swept by the vector $\langle a, b \rangle$ as it moves along the other vector $\langle c, d \rangle$. Let us, instead, move $\langle a, b \rangle$ in two simpler motions, as shown in figure 2.6. The area swept during the first move is ad, and during the second move $-bc$; the minus sign is due to the fact that the segment moves "backwards." The total area swept is thus $ad - bc$. It remains to observe that the area swept does not depend on the path of the moving vector $\langle a, b \rangle$, as long as it moves parallel to itself. Indeed, the rate of change of the area swept equals the length of the segment times the speed in the perpendicular direction. Thus the

---

[4]Some unfortunates, including the author, have been taught the latter formula as the definition but not its geometrical meaning.

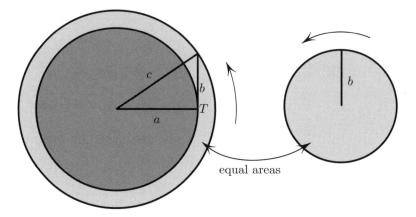

Figure 2.7. The proof of Pythagorean theorem by sweeping.

area swept is the length times the displacement in the perpendicular direction. In particular, how this displacement was achieved does not matter.

Problem. *Give a similar "sweeping" explanation of the formula of expansion of the determinant in a row*:

$$\begin{vmatrix} a_{11} & a_{12} & a_{13} \\ a_{21} & a_{22} & a_{23} \\ a_{31} & a_{32} & a_{33} \end{vmatrix} = a_{11} \begin{vmatrix} a_{22} & a_{23} \\ a_{32} & a_{33} \end{vmatrix} - a_{12} \begin{vmatrix} a_{21} & a_{23} \\ a_{31} & a_{33} \end{vmatrix} + a_{13} \begin{vmatrix} a_{21} & a_{22} \\ a_{31} & a_{32} \end{vmatrix}$$

Hint: Move the parallelogram Π formed by the last two row vectors in the direction of the x axis by a_{11}, then in the y direction by a_{12}, and finally in the z direction by a_{13}. Compare the volume swept with the volume swept by Π under the "diagonal" translation by $\langle a_{11}, a_{12}a_{13} \rangle$.

2.6 The Pythagorean Theorem by Rotation

Figure 2.7 shows a right triangle executing one full turn around an endpoint of its hypotenuse. The hypotenuse and the leg adjacent to the pivot sweep out disks, while the remaining leg sweeps out a ring.

We have

$$\pi a^2 + (\text{area of the ring}) = \pi c^2.$$

Proving the Pythagorean theorem amounts to showing that the area of the ring is πb^2. How do we prove this directly, without appealing to the theorem?

Here is a heuristic argument. The ring is swept by a moving segment of length b as the segment executes two simultaneous motions: sliding (in the direction of the segment) and rotating around the trailing point T of the segment. The key observation is this: *the sliding motion does not affect the rate at which the segment sweeps the area.* In other words, by subtracting the sliding velocity, and thus making the segment rotate in place around its trailing point, we do not affect the rate at which the segment sweeps area. This explains why the area of the ring equals the area of the disk in figure 2.7.

2.7 Still Water Runs Deep

A deceptively shallow statement can have deeper consequences. Here is an example of such a statement: "Barring external disturbance, still water in a container will remain still."[5] Actually, the obvious statement implies the following less obvious facts:

1. The Pythagorean theorem
2. Archimedes' law of buoyancy
3. The law of sines

The first of these is essentially explained by the previous "fish tank" argument; instead of the fish tank we could imagine a prism of water hanging in a large body of still water, as in figure 2.8. Since the prism is in equilibrium, the sum of torques (around any vertical edge) of the inward pressures on the vertical faces is zero. This zero torque condition is the same as (2.1), up to a sign, that is, it is the same as the Pythagorean theorem.

---

[5]This is again a special case of the law of conservation of energy, stating that the energy cannot be created. The more general the statement, the simpler it sounds.

Figure 2.8. The sum of torques on the imaginary prism of water is zero.

Achimedes' law. This can be proven in one stroke, as follows. The law states: *the buoyancy force acting on a submerged body (say a rock) equals the weight of the water displaced by the body.*

Proof. Imagine replacing the submerged rock with the identically shaped blob of water. This blob of water will hover in equilibrium, as mentioned above. The buoyancy on the water blob therefore equals the blob's weight. But the rock "feels" the same buoyancy since it has the same shape as the blob.   ◇

The law of sines. This law, we recall, states that for any triangle the length of each side is proportional to the sine of the opposite angle:

$$\frac{a}{\sin \alpha} = \frac{b}{\sin \beta} = \frac{c}{\sin \gamma}.$$

Proof. To prove this law using hydrostatics, imagine a thin endless tube in the form of triangle $\triangle ABC$, filled with water, placed in a vertical plane (figure 2.9). Alternatively, we can just imagine the triangular tube of water suspended in a surrounding body of water.

Let us position the side AB horizontally; the pressures at A and B are then equal, and $p_A - p_C = p_B - p_C$. But the pressure differences are proportional to the difference in depths: $p_A - p_C = kb \sin \alpha$ and $p_B - p_C = ka \sin \beta$ where k is the coefficient of proportionality. We conclude that $b \sin \alpha = a \sin \beta$. A similar argument shows that $c \sin \beta = b \sin \gamma$. The law of sines follows.   ◇

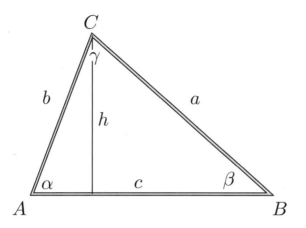

Figure 2.9. A thin water-filled tube of triangular shape used in the proof of the law of sines.

2.8 A Three-Dimensional Pythagorean Theorem

Theorem. *For any tetrahedron bounded by three mutually orthogonal planes and the fourth plane not parallel to any of the other three, one has*

$$a^2 + b^2 + c^2 = d^2, \tag{2.4}$$

where a, b, and c are the areas of the faces on the mutually orthogonal planes, and d is the area of the remaining face.

A physical proof. Fill our tetrahedron with compressed gas. The sum of all the internal pressure forces upon the pyramid is zero:

$$\mathbf{F}_a + \mathbf{F}_b + \mathbf{F}_c = -\mathbf{F}_d, \tag{2.5}$$

since otherwise our container would accelerate spontaneously in the direction of the resultant force, giving us a free source of energy in violation of the law of conservation of energy—a law which, to our knowledge, has so far been enforced with 100% compliance.

Since $(\mathbf{F}_a + \mathbf{F}_b) \perp \mathbf{F}_c$, the Pythagorean theorem yields

$$|\mathbf{F}_a + \mathbf{F}_b|^2 + |\mathbf{F}_c|^2 = |\mathbf{F}_d|^2.$$

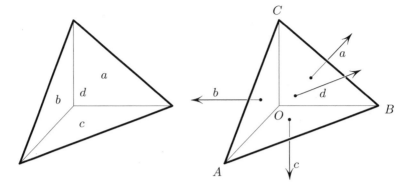

Figure 2.10. Three-dimensional version of the Pythagorean theorem: the areas satisfy (2.4).

Similarly,

$$|\mathbf{F}_a + \mathbf{F}_b|^2 = |\mathbf{F}_a|^2 + |\mathbf{F}_b|^2.$$

We conclude

$$|\mathbf{F}_a|^2 + |\mathbf{F}_b|^2 + |\mathbf{F}_c|^2 = |\mathbf{F}_d|^2. \tag{2.6}$$

Now $|\mathbf{F}_a| = \text{pressure} \cdot \text{area} = pa$; similarly, $|\mathbf{F}_b| = pb$, $|\mathbf{F}_c| = pc$, and $|\mathbf{F}_d| = pd$. Substituting into (2.6) and canceling p^2 gives (2.4).   ◇

To summarize: the area theorem (2.4) amounts to saying that the pressurized container of the shape shown in figure 2.10 provides zero thrust! A simple physical observation gives a neat mathematical theorem.

A mathematical "cleanup." A skeptic may complain about the lack of mathematical rigor in getting to (2.5). Indeed, we had appealed to the law of conservation of energy, which had not been given a precise mathematical formulation. To answer this complaint, we observe: (2.5) *is equivalent to the invariance of the volume of the pyramid under translations.*

Indeed, (2.5) is equivalent to saying that for any vector $\mathbf{r}$

$$\mathbf{F}_a \cdot \mathbf{r} + \mathbf{F}_b \cdot \mathbf{r} + \mathbf{F}_c \cdot \mathbf{r} = -\mathbf{F}_d \cdot \mathbf{r}.$$

But the term $\mathbf{F}_a \cdot \mathbf{r}$ gives the volume swept by the face OBC as it is translated by the vector $\mathbf{r}$, with a similar statement for the other faces. In short, the last equation expresses the fact as the pyramid is translated by $\mathbf{r}$, the volume gained by the faces a, b, c equals the volume lost by the face d.

Putting it differently, let $V = V(\mathbf{r}) = V(x, y, z)$ be the volume of the pyramid translated by $\mathbf{r} = \langle x, y, z \rangle$. Of course, V is independent of $\mathbf{r}$, that is, partial derivatives with respect to each of the three variables vanish:

$$\langle V_x, V_y, V_z \rangle \equiv \nabla V(\mathbf{r}) = \mathbf{0}.$$

Physically, the gradient vector $\nabla V(\mathbf{r})$—the vector of partial derivatives—is the resultant force of internal pressures of gas at pressure $p = 1$ on the container's walls.

2.9 A Surprising Equilibrium

Why does the Pythagorean theorem have so many different proofs? Perhaps because it is so basic. Even when we limit ourselves to physical, or physics-inspired proofs, there are several; one such proof was given in section 2.2, with two more to come. In preparation for one of these proofs we consider first a simple mechanism of independent interest. In the following section we will use this mechanism to prove the Pythagorean theorem (again!).

Problem.[6] *A small ring C slides without friction on a rigid semicircle. Two identical zero-length springs[7] C A and C B connect the ring to the diameter's ends. Prove:* the ring is in equilibrium in **any** position on the semicircle.

---

[6]I had stumbled upon this observation when thinking of the motion of a large artificial satellite.

[7]By the definition, the tension of a zero-length spring varies in direct proportion to its length. In particular, zero tension corresponds to zero length. The potential energy of such a spring is proportional to the square of its length (see section A.1).

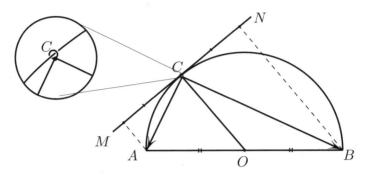

Figure 2.11. Proof by springs.

Proof. The ring is in equilibrium if the tangential components of all forces acting on the ring cancel each other. Three forces act on the ring: the normal reaction force from the circle and the two tension forces $\overline{CA}$ and $\overline{CB}$ (we chose Hooke's constant $k = 1$), seen in figure 2.11. Only the last two forces have nonzero tangential components, and we have to show that these two components cancel each other. To that end we just note that the projections of the two radii onto MN satisfy

$$\overline{OA}_{MN} = \overline{OB}_{MN},$$

and, since $OC \perp MN$, these radii have the same projections as the two forces:

$$\overline{OA}_{MN} = \overline{CA}_{MN}, \ \overline{OB}_{MN} = \overline{CB}_{MN}.$$

The projections of the two forces CA and CB cancel and the ring is in equilibrium (in any position).   ◇

2.10 Pythagorean Theorem by Springs

Having just shown that the ring in figure 2.11 is in equilibrium, we thereby proved the Pythagorean theorem. Indeed, since the ring is in equilibrium at any point C on the circle, it takes zero force, and thus zero work, to slide the ring C to A. This means that the potential energy did not change during sliding, so that the initial energy equals

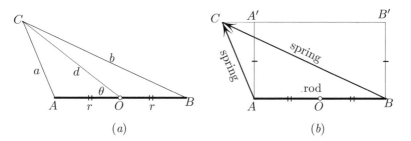

Figure 2.12. (a) $a^2 + b^2$ is independent of the angle θ; (b) the torques balance, since the components of the forces normal to AB are equal and so are the levers.

the final energy:

$$\frac{k}{2}a^2 + \frac{k}{2}b^2 = \frac{k}{2}0^2 + \frac{1}{2}c^2.$$

We used the fact that the potential energy of a zero-length spring of length x is $\frac{k}{2}x^2$, where k is a constant, (see section A.1). We conclude that $a^2 + b^2 = c^2$.

2.11 More Geometry with Springs

The ring-on-the-circle problem (section 2.9) can be reinterpreted in the following way, equally surprising, I think. The device in figure 2.12 is suggested by the sliding ring on a wire shown in figure 2.11. The difference in the present figure is that I put C in a fixed position in the plane, while allowing the segment AB to pivot on its midpoint O. In addition, the distance from C to O is now arbitrary. Two identical zero-length springs AC and BC compete, trying to rotate AB in opposite directions.

Problem A. *Prove that in the mechanism described above, the rod is in equilibrium in any orientation.*

Problem B. *Prove that for any triangle $\triangle ABC$*

$$a^2 + b^2 = 2(d^2 + r^2),$$

where $r = OA = OB$ *is half the length of the side* AB *and where* $d = OC$ (see figure 2.12).

Solutions. *Problem A*: Let us choose Hooke's constant $k = 1$ for our two springs. Then $\overline{AC}$ and $\overline{BC}$ are the forces upon the ends A and B in figure 2.12. The torques[8] of these two forces relative to the pivot O have equal magnitudes: indeed, the levers are equal, $OA = OB$, as are the two forces' normal components, $AA' = BB'$, in figure 2.12. These torques are opposing, so that the rod is in equilibrium. $\diamond$

Problem B: Since the rod is in neutral equilibrium[9] zero work is needed to aim the rod directly at the point C. This means that the potential energy of the rod in any position is the same as in this special one:

$$\frac{1}{2}a^2 + \frac{1}{2}b^2 = \frac{1}{2}(d-r)^2 + \frac{1}{2}(d+r)^2$$

or

$$a^2 + b^2 = 2(d^2 + r^2).$$ $\diamond$

2.12 A Kinetic Energy Proof: Pythagoras on Ice

Imagine standing in the corner of a perfectly frictionless "skating rink" (figure 2.13). Your shoes are perfectly frictionless. Pushing off of the x axis, you start sliding with speed a along the y axis. Your kinetic energy is $ma^2/2$. Now push off of the y axis, acquiring speed b in the x direction, thus gaining extra kinetic energy $mb^2/2$ (during the push, the friction with the y axis is assumed to be zero). Your kinetic energy after these two pushes is[10] $\frac{ma^2}{2} + \frac{mb^2}{2}$. On the other hand, your final speed is the hypotenuse c of the velocity triangle, and your kinetic energy is therefore given by $\frac{mc^2}{2}$. Thus

$$\frac{ma^2}{2} + \frac{mb^2}{2} = \frac{mc^2}{2},$$

or $a^2 + b^2 = c^2$.

---

[8]For the background on torque see section A.5.
[9]We say that an equilibrium is *neutral* if any position is an equilibrium.
[10]Kinetic energy is a scalar and thus adds *arithmetically*.

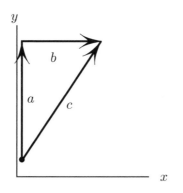

Figure 2.13. Kinetic energy after two consecutive pushes: $\frac{ma^2}{2} + \frac{mb^2}{2} = \frac{mc^2}{2}$.

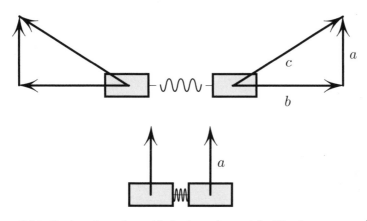

Figure 2.14. Cutting the string adds horizontal speed b. Kinetic energy mc^2 is acquired in two portions, first ma^2 and then mb^2.

2.13 Pythagoras and Einstein?

Here is a "cutting the string" proof of the Pythagorean theorem. It is essentially the same proof as the preceding one, just recast into a different form.

Let us compress a spring between two equal masses so that if released, the masses will fly apart each with speed b. We then tie the two masses together with a string to keep the spring compressed, as shown in figure 2.14.

Let us throw the "loaded" system with speed a as shown in figure 2.14,[11] and then, once the system is flying, we snip the rope, thus releasing the compressed spring. The resulting speed c of each mass is the hypotenuse of the velocity triangle with legs a, b. On the one hand, the kinetic energy of both masses is now $2 \cdot \frac{mc^2}{2} = mc^2$. But this energy was acquired in two portions: first, $2 \cdot \frac{ma^2}{2} = ma^2$ from the initial push, and second, $2 \cdot \frac{mb^2}{2} = mb^2$ from the spring. Thus

$$mc^2 = ma^2 + mb^2.$$

Cancellation of m gives the Pythagorean theorem.[12]

---

[11] We assume no gravity.

[12] Our facetious reference to Einstein is due to the fact that the system has the energy $E = mc^2$.

3

MINIMA AND MAXIMA

Max/min problems tend to be well suited for the physical approach. The reason for this is perhaps the fact many physical systems find maxima or minima automatically: a pendulum finds the minimum of potential energy; the light from a pebble on the bottom of the pool to my retina chooses the path of least time; a soap bubble chooses the shape of least volume; a chain hanging by two ends chooses the shape of lowest center of mass, and so on—the list is endless.

Here is a common pattern in finding a physical solution. Let us say we have to minimize a function. The main step is to invent a mechanical system whose potential energy is the given function. The minimum of the function corresponds to the minimum of the potential energy, which in turn corresponds to the equilibrium. The equilibrium condition, when written down, often already is in the form of a nice answer. In effect, we are inventing a mechanical "analog computer" which solves the problem by itself—we just need to read off the answer.

Here is a schematic representation of the correspondence between calculus and mechanics, for the case of a function of one variable x:

Calculus	**Physical interpretation**
The function $f(x)$	Potential energy $P(x)$
The derivative $f'(x)$	The force $F(x) = -P'(x)$
$f(x)$ minimal $\Rightarrow f'(x) = 0$	$P(x)$ is minimal $\Rightarrow F(x) = 0$ (equilibrium)

A note on the background. Precalculus and some very basic geometry should be enough to understand this chapter. Nevertheless, the physical approach lets us solve quite a few calculus problems, even some from multivariable calculus!

The physical background used in this chapter is described in the appendix. We use mechanical models of mathematical objects. These models consist of idealized elastic springs, ropes, sliding rings, compressed gas, and vacuum. All these imaginary objects are described in the short appendix, where the concepts of equilibrium, torque, and the centroid are explained as well.

The lunch is not quite free. Some of the problems here become one-liners when physics is employed, instead of being one-pagers when calculus is used. However, by the law of conservation of difficulty, this does not come free. The difficulty is shifted from making dull algebraic manipulations to that of inventing the right mechanical system.

Some highlights. The topics of this chapter include

1. An optical property of ellipses.
2. The line of best fit using springs.
3. Pyramids of least volume and centroids.
4. Maximal and minimal area problems.
5. Minimal surface area problems.
6. The inscribed angle theorem using mechanics.
7. Saving a drowning victim using weights.

Many if not all of these are calculus problems, but we solve them here without calculus.

3.1 The Optical Property of Ellipses

The ellipse is a kind of a "circle with two centers": one ties a string between two nails (F_1 and F_2) and moves the pencil to keep the string taut; the pencil will trace an ellipse. To be precise, the ellipse consists of all points for each of which the sum of distances to two given points (called the foci) is a given constant.

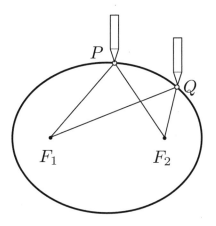

Figure 3.1. The definition of the ellipse: $PF_1 + PF_2 = \text{constant}$.

An ellipse has this remarkable property: a ray of light emitted from one focus will, upon reflection from the ellipse, pass through the other focus—this is true for any direction of the emitted ray. Playing laser tag in an elliptical room with reflecting wall would be a lot of fun.

Alternatively, imagine playing squash in an elliptical room; standing at one focus F_1 and throwing the ball, I will hit the person standing at the other focus F_2 no matter how bad my aim is (assuming the ball bounces so that the incidence and reflection angles are equal). Of course, if the person at F_2 ducks, then the ball will pass F_2 and will hit me after one more bounce off the wall.

What is the explanation of this remarkable property? Here is a precise geometrical statement of the problem, followed by the answer.

Problem. *Let P be a point on an ellipse with the foci F_1 and F_2, and let MN be the tangent at P, figure 3.2a. Prove that*

$$\angle F_1 PM = \angle F_2 PN.$$

Solution. How might we prove this property? A brute force solution is to (i) write the equation of an ellipse, (ii) compute the two angles in question, and (iii) verify that the expressions are equal. This

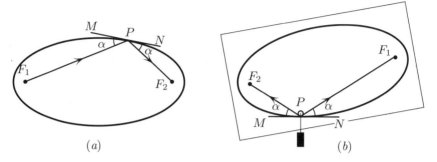

Figure 3.2. (*a*) The optical property; (*b*) a mechanical proof.

approach can lead to a finger-breaking calculation and, to add insult to injury, little understanding would be gained. Instead, a direct approach shown here is much shorter and shows "what's going on."

Let us hang a string from two nails F_1, F_2 driven into a board (figure 3.2(*b*)), letting a weighted pulley roll on the string as shown. If we move the pulley left or right, while keeping the string taut, the pulley will trace out an ellipse. By orienting the board appropriately, we can arrange for an arbitrary point P on the ellipse to be the lowest point on the ellipse, as in figure 3.2(*b*), so that the tangent line becomes MN horizontal. But the string forms equal angles with the horizontal. Indeed, the three forces acting on the pulley (the two tensions and the weight) are in balance; in particular, the net horizontal force is zero:

$$T_1 \cos \alpha_1 - T_2 \cos \alpha_2 = 0,$$

where T_1, T_2 are the tensions in the two straight parts of the string. But $T_1 = T_2$ since the pulley is frictionless,[1] and we get $\alpha_1 = \alpha_2$. ◇

Heron's principle. Heron's principle states that light takes the shortest path. By the definition of the ellipse, every path $F_1 P F_2$ has the same length; it sounds silly but is true that every such path is a shortest way from F_1 to the ellipse to F_2. Thus $F_1 P F_2$ is a path of

---

[1] Indeed, any difference in tensions would cause angular acceleration of the pulley in the direction of the greater tension.

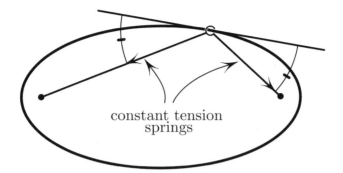

Figure 3.3. The ring is in equilibrium, hence the angles are equal.

some ray. On the other hand, we showed that $\angle F_1 PM = \angle F_2 PM$. In other words, we showed that Heron's principle is consistent with the equality of the angles of incidence and reflection.

3.2 More about the Optical Property

Here is a slightly different mechanical proof of the optical property. Consider a ring sliding frictionlessly along the ellipse, and use two constant tension $T = 1$ springs[2] to attach the ring to each of the foci. The potential energy of our mechanical system equals the total length of the springs, which, by the definition of the ellipse, is constant. The ring is therefore in equilibrium at any location, and the tangential components of the forces upon the ring are therefore in balance: $T \cos \alpha_1 = T \cos \alpha_2$, implying $\alpha_1 = \alpha_2$.

3.3 Linear Regression (The Best Fit) via Springs

Imagine a collection of data points (x_i, y_i) in the plane. We are asked to find the straight line $y = ax + b$ that best fits this set of data. What does "best" mean? To answer this, for each x_i we think of $y = ax_i + b$

---

[2] For the description of these, see section A.1.

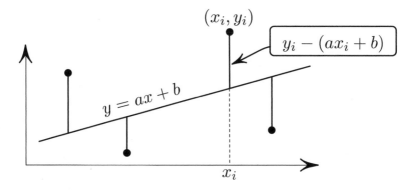

Figure 3.4. Which line minimizes the sum of errors (3.1)?

as the predicted value, while y_i is the observed or measured value. The mismatch between these two values is $y_i - (ax_i + b)$, called the error (figure 3.4). "Best line" here means the line for which the sum of squares of the errors is minimal. The precise formulation of the problem of best fit, also called the problem of *linear regression*, follows.

Problem. *Given N data points (x_k, y_k) in the plane, find the straight line $y = ax + b$ which fits these data best in the sense of minimizing the sum of squares of errors*

$$S(a, b) = \sum_{i=1}^{N}(y_i - (ax_i + b))^2. \qquad (3.1)$$

The unknowns in this problem are the slope a and the intercept b of the "best" straight line. The standard method to find the minimum of (3.1) is to set the partial derivatives with respect to a and b to zero. Here is a mechanical shortcut to the answer.

Solution. The unknown straight line is to be imagined as a rigid rod (figure 3.5). Let us pass the rod through frictionless sleeves constrained to vertical lines $x = x_i$ by frictionless guides. Each sleeve is connected to a nail (hammered into a data point) by a

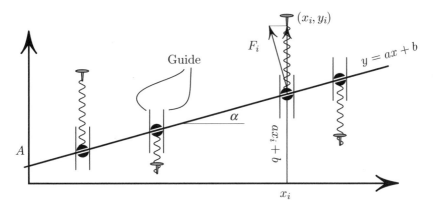

Figure 3.5. Sum of squares = potential energy (3.1) is minimized when the equilibrium conditions (3.2) on forces and torques hold.

zero-length spring.[3] Let us take Hooke's constant to equal 2, so that the potential energy of each spring is simply the square of its length. The sum (3.1) has now acquired a physical meaning of potential energy!

If the sum of squares is minimal, then the potential energy of our mechanical system is minimal, and consequently the rod is in equilibrium. The only forces the rod "feels" are the normal reactions F_i from the sleeves; the sum of these forces vanishes, as does the sum of their torques[4] relative to the point of intercept A:

$$\sum_{i=1}^{N} F_i = 0, \quad \sum_{i=1}^{N} d_i \, F_i = 0, \tag{3.2}$$

where d_i is the distance from the intercept to the sleeve. Note that $d_i \cos \alpha = x_i$. Now to get an expression for F_i, consider the balance of forces upon the sleeve. The sleeve feels (i) the reaction force $-F_i$ from the rod, (ii) the pull of the spring, $y_i - (ax_i + b)$, and (iii) the reaction from the guide in the x direction. Only two of these forces have nonzero y components, and they are in

---

[3]The background on zero-length springs is in section A.1. For our purposes here it suffices to recall that the potential energy of a zero-length spring of Hooke's constant k is $\frac{k}{2}x^2$.

[4]For the background see sections A.5 and A.6.

balance: $F_i \cos \alpha = y_i - (ax_i + b)$. Using these expressions for d_i and F_i in (3.2) we obtain

$$\begin{cases} \sum y - a \sum x - Nb = 0 \\ \sum xy - a \sum x^2 - b \sum x = 0. \end{cases} \tag{3.3}$$

This is a system of two equations with two unknowns a and b which, when solved, produces the "best" slope and intercept.   ◇

Note that the same result (3.3) can be obtained directly by setting partial derivatives of the error in (3.1) to zero:

$$\frac{\partial S}{\partial b} = 2 \sum (y_i - (ax_i + b)) = 0$$

and

$$\frac{\partial S}{\partial a} = 2 \sum x_i(y_i - (ax_i + b)) = 0.$$

Now we have a physical interpretation of these conditions: the first expresses the vanishing of the net force on the rod, while the second expresses the vanishing of the torque relative to the point of intercept A.

3.4 The Polygon of Least Area

The following is a good example where a physical approach works very well, leading to a very quick solution. This problem represents a whole class of similar problems.

Theorem. *Consider an n-sided polygon P of least possible area circumscribed around a given closed convex curve[5] K. Each tangency point of K with a side of P is the midpoint of that side.*

Proof. The figure shows three straight lines enclosing the curve K, with a vacuum inside the triangle and with gas of pressure $p = 1$ outside pressing on the lines (impenetrable by the gas). The obstacle

---

[5] K is not given by a formula—all we know is that it is convex.

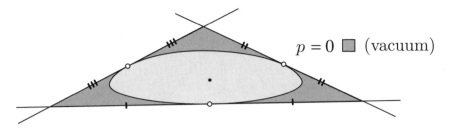

$p = 0$ ■ (vacuum)

Figure 3.6. The three rods enclosing a vacuum bubble are pressed inward against the curve K by the outside gas.

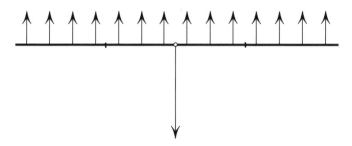

Figure 3.7. If a force uniformly distributed over a segment is balanced by a force applied at a segment's point, then that point is the midpoint.

K prevents the triangle from collapsing to a point. Speaking intuitively, the triangle tries to do the "next best thing" to collapsing: it tries to minimize its area. More precisely, let the rods form a triangle of least area A; the rods are not connected at the vertices. I claim that each rod is then in equilibrium. Indeed, the potential energy of our system equals[6] the area of the vacuum times the pressure p, so that

minimal area ≡ minimal potential energy ≡ equilibrium.

Since each rod is in equilibrium, the outward pressure on the rod at the point of contact balances the inward pressure of the gas. This implies (see figure 3.7) that the point pressure is applied at the midpoint; otherwise the torque of all the forces upon the rod around

[6]For justification see section A.4.

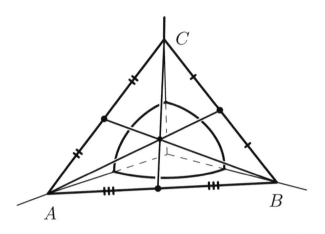

Figure 3.8. If the tangent plane minimizes the volume of the pyramid, then the point of tangency is the centroid of $\triangle ABC$.

the tangency point will be nonzero. Simply speaking, the rod would have rotated if the contact point were not the midpoint.

3.5 The Pyramid of Least Volume

Problem. *Of all planes tangent to the ellipsoid $x^2/a^2 + y^2/b^2 + z^2/c^2 = 1$, one of them cuts the pyramid of least possible volume from the first octant $x \geq 0$, $y \geq 0$, $z \geq 0$. Show that the point of tangency of that plane is the centroid[7] of the face ABC (figure 3.8).*

Solution by physics. Let us imagine the tangent plane separating the vacuum inside the pyramid from the air outside in the first octant. The air presses only on that part of the plane which is in the first octant. The plane passes the coordinate planes without resistance, in our thought experiment. The air pushes the plane against the rigid ellipsoid, trying to "mash" the volume of the pyramid to a minimum. With the minimal volume the plane will be in equilibrium—intuitively, any motion which expands the vacuum "bubble" would

---

[7]The centroid, or the balance point of a triangle, is the intersection of its medians.

require effort. More precisely, the potential energy of the system is proportional to the volume of the vacuum (section A.4), so that minimal volume implies minimal energy, which in turn implies an equilibrium for the rolling plane. But this is the balance between the uniformly distributed air pressure on $\triangle ABC$ from the outside and a point pressure from the ellipsoid on the other. *By the definition, this means that the contact point is the centroid of the triangle $\triangle ABC$.* The centroid of a triangle is the point of intersection of its medians, as explained in section 3.15. ◇

Just for comparison, here is a conventional solution.

Solution by calculus (not necessarily to be read, but for the length comparison). We start by expressing the volume of the pyramid in terms of the point of tangency (x_0, y_0, z_0). The normal vector to the ellipsoid at this point is

$$\mathbf{N} = \nabla \left(\frac{x^2}{a^2} + \frac{y^2}{b^2} + \frac{z^2}{c^2} \right) = 2 \left\langle \frac{x_0}{a^2}, \frac{y_0}{b^2}, \frac{z_0}{c^2} \right\rangle,$$

and the equation of the tangent plane is

$$(\mathbf{r} - \mathbf{r}_0) \cdot \mathbf{N} = 0,$$

where $\mathbf{r}_0 = \langle x_0, y_0, z_0 \rangle$, or

$$\frac{(x - x_0)x_0}{a^2} + \frac{(y - y_0)y_0}{b^2} + \frac{(z - z_0)z_0}{c^2} = 0.$$

With $\frac{x_0^2}{a^2} + \frac{y_0^2}{b^2} + \frac{z_0^2}{c^2} = 1$, the equation of the tangent plane simplifies to

$$\frac{x_0 x}{a^2} + \frac{y_0 y}{b^2} + \frac{z_0 z}{c^2} = 1.$$

We compute the x intercept X by setting $y = z = 0$, and similarly for the other two intercepts:

$$X = \frac{a^2}{x_0}, Y = \frac{b^2}{y_0}, Z = \frac{c^2}{z_0}.$$

The volume of the pyramid is now expressed in terms of the point of tangency:

$$V = \frac{1}{3}\left(\frac{1}{2}XY\right)Z = \frac{1}{6}XYZ = \frac{(abc)^2}{x_0 y_0 z_0}.$$

We have to minimize V, that is, to maximize $x_0 y_0 z_0 \equiv f(x_0, y_0, z_0)$ subject to the constraint

$$g(x_0, y_0, z_0) \equiv \frac{x_0^2}{a^2} + \frac{y_0^2}{b^2} + \frac{z_0^2}{c^2} = 1.$$

For the minimality of f we must have $\nabla f = \lambda \nabla g$, where λ is the Lagrange multiplier; more explicitly, this gives

$$\begin{cases} y_0 z_0 = 2\lambda \frac{x_0}{a^2} \\ x_0 y_0 = 2\lambda \frac{y_0}{b^2} \\ x_0 y_0 = 2\lambda \frac{z_0}{c^2}. \end{cases} \qquad (3.4)$$

Multiplying the first equation by x_0, the second by y_0, and the third by z_0 we make all the left-hand sides equal to each other. The resulting right-hand sides are therefore equal as well:

$$\frac{x_0^2}{a^2} = \frac{y_0^2}{b^2} = \frac{z_0^2}{c^2}.$$

Finally, from $g(x_0, y_0, z_0) = 1$ we conclude that

$$\frac{x_0^2}{a^2} = \frac{y_0^2}{b^2} = \frac{z_0^2}{c^2} = \frac{1}{3},$$

$$x_0 = \frac{a}{\sqrt{3}},$$

and

$$X = \frac{a^2}{x_0} = a\sqrt{3}.$$

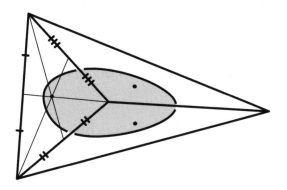

Figure 3.9. If the volume of a pyramid circumscribed around a convex body is minimal, then the face's centroids are the tangency points.

We observe that $x_0 = \frac{1}{3}X$; similarly, $y_0 = \frac{1}{3}Y$ and $z_0 = \frac{1}{3}Z$. In other words,

$$(x_0, y_0, z_0) = \frac{1}{3}\left((X, 0, 0) + (0, Y, 0) + (0, 0, Z)\right), \qquad (3.5)$$

that is, the tangency point is indeed the centroid of the triangle ABC. ◇

3.6 A Theorem on Centroids

This result has been known for some time (see [D]). This problem is a generalization of the preceding one. I am stating it here because this generalization is so elegant it deserves its own space.

Theorem. *Let K be a convex body in $\mathbb{R}^3$, and let $ABCD$ be a tetrahedron (i.e., a triangular pyramid) of least possible volume containing K. Then the point of tangency of each face with K is that face's centroid.*

Proof. We imagine four planes enclosing K and bounding a pyramid-shaped bubble of vacuum. The planes can pass through each other without interaction, but they cannot penetrate K. The air pressure outside the bubble of vacuum forces the planes to press against K. The volume of the vacuum pyramid is in direct proportion to the potential energy.[8] Hence if the pyramid has minimal volume, then the potential energy will be minimal, and hence all planes will be in equilibrium. Thus the outward point pressure on each face at the tangency point balances the uniformly distributed inward pressure. This implies that the tangency point is the centroid of the face. Indeed, we can think of the inward pressure as gravitational force, uniformly distributed over the triangle, and then the tangency point is the point of balance—the centroid!   ◇

3.7 An Isoperimetric Problem

Here is a nonstandard solution of a standard calculus problem. As a side benefit, this solution actually answers many more questions than just this one, as shown in the following section.

Problem. *Find the dimensions of a circle and a square of given combined perimeter with smallest combined area.*

Answer. $x = d$: the circle is inscribable in the square.

Solution. Our mechanical system consists of a rope forming a loop; part of this loop is wrapped around a square and part around a circle, with a neck passing through a tube as shown in figure 3.11.

On one side of the tube, the rope is kept in a square shape by a constraining mechanism as shown in figure 3.11.

Imagine now gas enclosed by the loop as shown in figure 3.11. The gas tries to expand but cannot do so indefinitely because the rope is inextensible. The potential energy of the gas is a decreasing function of the area.[9] The minimal area therefore corresponds to maximal potential energy and thus to an equilibrium[10] of the system.

---

[8] This is plausible intuitively; the proof is given in the appendix (see section A.4).

[9] Indeed, if we let the gas expand, it will do positive work and thus *lose* that same amount of its potential energy.

[10] An unstable one, since the energy is maximal, but so what?

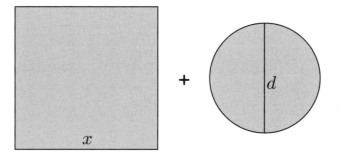

Figure 3.10. Minimize the area given the combined perimeter L.

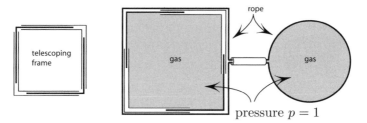

Figure 3.11. The two containers try to expand against the rope's tension.

Therefore (and this is the key point), *the tension of the rope is the same everywhere in the rope; in particular in the square and in the circular parts:*

$$T_1 = T_2. \tag{3.6}$$

In the next paragraph I will show that $2T_1 = px$ and $2T_2 = pd$, and thus $x = d$: the square's side equals the circle's diameter, as claimed.

The tension of the rope. First consider a square container enclosing compressed two-dimensional gas with pressure p. Let T_1 denote the tension of the side of the square.[11] Figure 3.12 shows the force px of the gas pushing a side of the square outward, balancing the combined force $2T_1$ of the tension of the rope.

---

[11]To avoid misunderstandings, by tension I mean the force that needs to be applied to keep each of the two ends of the rope from separating if the rope is cut.

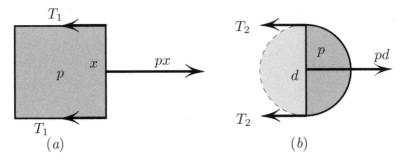

Figure 3.12. (a) The gas pushes the right side of the square outward to the right (px); the neighboring sides pull that side with the combined force $2T_1$ to the left. (b) The right semidisk feels pressure on its diameter, and the opposing pull by the tension of the rope.

The force balance gives

$$2T_1 = px. \tag{3.7}$$

Similarly for the circle, we consider the force balance on a semidisk: the diameter is pushed to the right, and the rope is pulled left; the balance yields

$$2T_2 = pd. \tag{3.8}$$

Using $T_1 = T_2$ we conclude:

$$x = d;$$

this solves the problem. A very compact answer!   ◇

Just for comparison with a physical solution, here is a conventional one.

A conventional solution. The combined perimeter is fixed:

$$4x + \pi d = L,$$

so that

$$d = \frac{L - 4x}{\pi}. \tag{3.9}$$

The combined area is therefore

$$A(x) = x^2 + \frac{\pi d^2}{4} = x^2 + \frac{\pi}{4} \left(\frac{L - 4x}{\pi} \right)^2.$$

Differentiating by x, we get

$$A'(x) = 2x + \frac{\pi}{4} \cdot 2 \cdot \left(\frac{L - 4x}{\pi} \right) \left(\frac{-4}{\pi} \right) = 2x - 2\frac{L - 4x}{\pi}.$$

From $A'(x) = 0$ we obtain

$$x - \frac{L - 4x}{\pi} = 0,$$

or

$$x = \frac{L}{\pi + 4}.$$

Substituting into (3.9) and simplifying we obtain

$$d = \frac{L}{\pi + 4}.$$

In particular, $x = d$.   ◇

The mechanical approach gives a direct physical meaning to $A'(x) = 0$—namely, the tension of the rope is the same throughout. Moreover, a curious feature of the physical approach is that we never had to write the expression for $A(x)$ (the very quantity we are minimizing), and never had to differentiate (although one could argue that by writing the condition $T_1 = T_2$ we actually differentiated, just in different terms). With the physical approach we go almost directly to the answer, once the mechanical model has been set up!

The next three examples show how easy it is to solve related problems once the mechanical model has been invented.

Problem. *Find the dimensions of a rectangle and an equilateral triangle of a given combined perimeter and of smallest combined area.*

Problem. *Find the dimensions of a regular n-gon and a regular m-gon of a given combined perimeter and of smallest combined area.*

Problem. *A circle, a square, and an equilateral triangle have a prescribed combined perimeter. Find the relative dimensions of the three figures which minimize the total area.*

Solution. We give a quick solution to the third problem. Reasoning just as above, we conclude that the minimal area condition, that is, the equilibrium condition, requires the tension of the boundary rope to be the same at each of the shapes:

$$T_{\text{circle}} = T_{\text{square}} = T_{\text{triangle}}. \tag{3.10}$$

By picking the gas pressure $p = 1$, we obtain the following tensions from the force balance equations:

$$T_{\text{circle}} = \frac{x}{2}, T_{\text{square}} = \frac{y}{2}, T_{\text{triangle}} = \frac{z\sqrt{3}}{6},$$

where x is the diameter of the circle, y is the side of the square, and z is the side of the triangle. The derivation for the triangle is shown in figure 3.13; for the circle and the square it was done on page 41. Equality of tensions (3.10) gives two equations

$$x = y = z\frac{\sqrt{3}}{3},$$

which specify the desired proportions of the three sizes.   ◇

To solve this problem by calculus, we would have to minimize the area as a function of three variables with a constraint—a subject usually covered in multivariable calculus.

3.8 The Cheapest Can

Every calculus student is exposed to a set of dreaded "can" problems in which the manufacturer wants to save money by producing cans of most efficient proportions. In this section I describe a physical

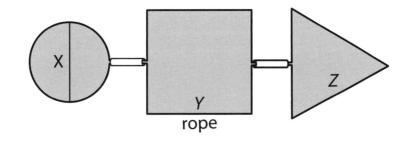

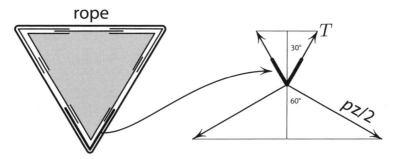

Figure 3.13. Balancing the forces on the corner bracket gives $2T \cos 30° = 2(pz/2)$
$\cos 60°$, or $T = \frac{z\sqrt{3}}{6}$ if $p = 1$.

solution of such a problem; in the next section I show how this
solution can be adapted to a wider class of problems.

Problem. *What are the proportions of a cylindrical can of minimal
total surface area containing a given volume?*

Solution. Let us imagine a cylindrical can constructed so that it can
telescope freely so as to change both its height and its radius. One
should imagine a mechanism built as suggested in figure 3.11 in the
previous problem. Let us now fill the can with water of volume V,
and enclose it in a film of constant surface tension[12] σ. One can
imagine soap film that doesn't burst and that slides without friction
along anything it touches.

---

[12] See section A.2 for a quick explanation of the concept of surface tension with all the
necessary background.

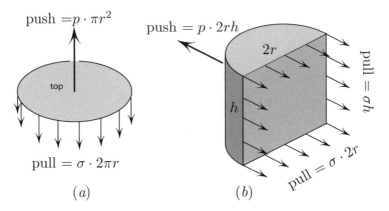

Figure 3.14. The balance between the surface tension and the pressure: (*a*) the top of the cylinder and (*b*) the half-cylinder.

The potential energy E of the film whose surface tension is a constant σ is proportional to the area: $E = \sigma A$ (see section A.2). Thus, the minimal-area shape is also the least-energy shape, and hence an equilibrium shape. In particular, the downward pull on the can's top (figure 3.14), acting along the circle, balances the upward pressure of the enclosed water upon the top:

$$\sigma \cdot 2\pi r = p \cdot \pi r^2. \qquad (3.11)$$

Moreover, the half-cylinder (the skin + the water) shown in figure 3.14 is pulled by the surface tension by the adjacent film and is pushed by the fluid upon the rectangular face:

$$\sigma \cdot (4r + 2h) = p \cdot 2rh.$$

Dividing this equation by (3.11) and simplifying, we obtain

$$h = 2r.$$

In other words, the "best" cylinder is the one whose axial cross section is a square.   ◇

3.9 The Cheapest Pot

Can mechanics handle a generalization of the "cheapest can" problem, where the costs of the can's top, sides, and bottom differ from each other? For instance, what are the dimensions of the cheapest topless can? With the addition of just one sentence, the solution of the previous problem carries over almost verbatim. Here are the details.

Problem. *The top, the sides, and the bottom of a cylindrical can are to be made of different materials with respective costs of a, b, and c cents per unit area. Find the proportions of the cheapest can of fixed volume.*

Solution.

The mechanical device. Just as in the preceding problem, we start with a telescoping cylinder-shaped shell whose radius and height can be changed freely. Let us fill this cylinder with water. Now, however, we will use three different films[13] with surface tensions a, b, and c equal to the respective costs. We stretch a patch of film with surface tension a and glue its edges to the top circle of the cylinder, so that the film lies along the top disk. We next wrap a film of constant tension b around the wall of the can, gluing it to the top and the bottom circles. The film will try to squeeze the side of the cylinder and will also pull the top and the the bottom circles toward each other. Finally, we repeat for the bottom disk what we did for the top, using the film c.

Finishing up. The three films try to squeeze the can. The total cost is the weighted sum of the areas of the top, side, and bottom: $\pi r^2 \cdot a + 2\pi rh \cdot b + \pi r^2 \cdot c$, which is precisely the potential energy of our mechanical system;[14] the cheapest can therefore is in equilibrium. In particular, the top of the can is pulled down by the film hugging the side, and pushed up by the internal pressure of the

---

[13]We again refer to the background on films, section A.2.

[14]Since (potential energy of the film) = (surface tension times) × (area); see section A.2 for details.

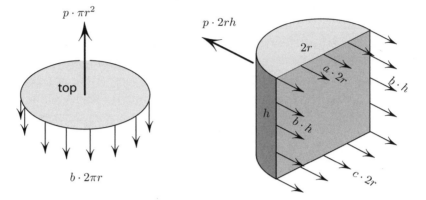

Figure 3.15. Minimizing cost with different materials for the top, sides, and bottom.

water (figure 3.15); these two forces are in balance:

$$b \cdot 2\pi r = p \cdot \pi r^2.$$

The half-cylinder in figure 3.15 is pulled by the surface tension in one direction and is pushed by the pressure p applied over the rectangular cross section $2r \times h$ in the opposite direction; these forces are in balance as well:

$$a \cdot 2r + b \cdot 2h + c \cdot 2r = p \cdot (2rh).$$

Dividing this equation by the preceding one and canceling terms we obtain the answer:

$$h = r\frac{a+c}{b}.$$

For a topless can made of uniform material where $a = 0$, $b = c$, this gives $h = r$. For the case $a = b = c$ the above formula gives $h = 2r$, in agreement with the result of the preceding problem.   ◇

3.10 The Best Spot in a Drive-In Theater

Problem. *From where can one see the movie screen at the largest angle? More precisely, given a vertical segment AB (the screen) and*

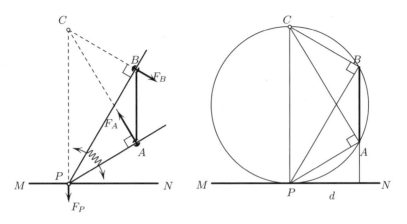

Figure 3.16. The hinge P slides freely along MN. The rays PA and PB slide freely through the sleeves at A and B. When the angle is maximal the system is in equilibrium. Hence the three forces acting upon the "jaw" APB are concurrent.

a horizontal line MN not intersecting AB, find the point P on MN such that $\angle APB$ *is maximal.*

The mechanical "computer." Consider a "jaw"—two rods connected by a hinge with a spring which tries to spread the rods apart. The rods are passed through the sleeves A and B (one at the top and the other at the bottom of the screen) so that they can slide through without friction. The jaw's apex P is welded to a small ring slipped around MN and sliding along MN without friction.

Solution.

An implicit answer. The longer the spring—that is, the larger the angle APB—the less the potential energy is. The maximal angle therefore corresponds to the minimal potential energy, that is, to the equilibrium of our system. But the jaw APB feels precisely three forces: F_P, F_A, and F_B; each force is perpendicular to the corresponding line, as shown in figure 3.16, since all contacts are frictionless. Since these forces have zero torque, their lines are concurrent, according to the lemma on concurrent forces in section A.6. *Consequently, for the largest possible angle APB the three perpendiculars to the lines MN, PA, and PB at the points*

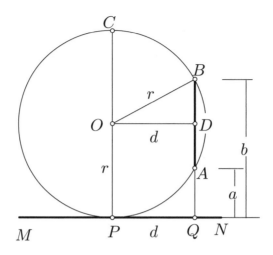

Figure 3.17. The optimal distance $d = \sqrt{ab}$.

P, A, and B are concurrent. We found a characteristic property of the desired configuration and thus have solved the problem, at least implicitly.   ◇

A constructive answer. To make the solution more explicit, note that $PA \perp AC$ and $PB \perp BC$ imply that A and B lie on the circle with the diameter PC. This circle is also tangent to MN since $PC \perp MN$. Thus *the optimal point P is the tangency point between the line MN and a circle passing through A and B.* There is precisely one such circle. The circle is easy to construct explicitly: its center is at the intersection of the perpendicular bisector of AB with vertical line through P.

The best distance is the geometric mean of the heights, $d = \sqrt{ab}$, where a and b are the heights of A and B. Indeed, referring to figure 3.17, we have

$$d = PQ = OD \overset{\Delta OBD}{=} \sqrt{OB^2 - BD^2},$$

where

$$OB = OP = DQ = \frac{a+b}{2}$$

and

$$BD = \frac{b-a}{2}.$$

Substituting the last two expressions into the first, we obtain

$$d = \frac{1}{2}\sqrt{(a+b)^2 - (a-b)^2} = \sqrt{ab}.$$

The best spot at a drive-in is given by the geometric mean of the top and the bottom heights of the screen! ◇

3.11 The Inscribed Angle

Theorem. *An inscribed angle in a circle depends only on the subtended arc and not on the location of the angle's vertex.*

Proof.

The mechanical device. For the proof, let us use the "jaw" (used in section 3.10) formed by two rods joined by a hinge P with a compressed spring trying to spread the rods apart. We constrain the hinge P to the circle, free to slide without friction. Each rod is slipped through a frictionless freely rotating sleeve attached at the ends A and B of the given arc, as shown in figure 3.18.

I claim that the system is in equilibrium for *any* location of P on the circle. To that end consider first the force on one rod (say, PB) *when P is held fixed artificially.* The rod is subject to the torque T around P and to the torque $F_B \cdot PB$ due to the reaction force F_B at the sleeve. The balance of torques around P gives $T = F_B \cdot PB$, which determines the value $F_B = T/PB$. Now let F be the tangential component of the force at P applied to the rod. This component balances the parallel to it component of the force at B (see figure 3.18(b)):

$$F = F_B \cos\theta = \frac{T\cos\theta}{PB} = \frac{T}{PQ}.$$

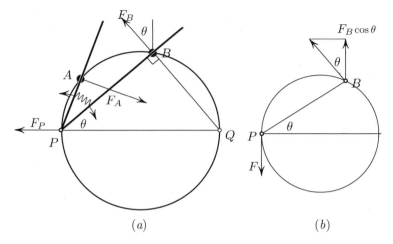

Figure 3.18. (*a*) The "analog computer." (*b*) Computing the force F necessary to keep P fixed.

We found that the tangent force F on the rod does not depend on the position of the rod. This means that the two rods push at the hinge at P in opposite directions with equal forces. We conclude that the hinge P is in equilibrium, as claimed.   ◇

3.12 Fermat's Principle and Snell's Law

A quick summary. Fermat's principle states that *the light traveling between two points "chooses" the path of least time*—at least, if the two points are sufficiently close to each other. To be more precise, any ray of light has the following time-minimizing property: for any two sufficiently close points A and B on the ray, the ray's arc AB gives the shortest possible travel time from A to B.[15]

---

[15] What really matters is not that A and B are close, but that a narrow fan of rays emanating from A does not focus before reaching B. The simplest illustration of this condition is offered by rays propagating on the sphere (figure 3.19). The points A and B are connected by two possible rays, AmB and AnB, the first of which is minimal; accordingly, the fan of rays from A does not focus on the least-time arc AmB. By contrast, the ray AnB is not the quickest way from A to B; this arc contains the focal point A' of the fan of rays from A. For the general theory associated with this problem I refer the reader to Milnor's book [M].

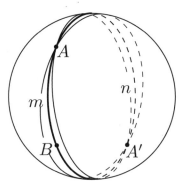

Figure 3.19. The arc AmB is the least-time path; the light from A does not focus on this arc. By contrast, the arc AnB is of extremal, but not shortest time; this property goes together with the existence of a focusing point on this arc.

Observing Fermat's principle in action. Fermat's principle predicts, among other things, the bending of rays in lenses as shown in figure 3.20. To minimize time, the light avoids the thicker part of the lens since traveling through glass takes longer.[16] This explains why the concave lens defocuses, while the convex lens focuses.

Another example in figure 3.21 shows a ray from the bottom of a pool to the eye. The path shown in the figure is quicker than the straight path, since it "pays" to shorten the "expensive" underwater part where light is slower.

How does light "know" which path is quickest? Light's ability to choose the quickest path seems magical. Doesn't picking the quickest path require information on other paths? As it actually turns out, light does know about other paths.[17] In fact, light travels along all possible paths, but, loosely speaking, the contributions of those cancel each other except for the shortest path and its near neighbors, whose contributions arrive "in sync."

---

[16]The ratio n of the speed of light in the vacuum to that in the glass is called the **index of refraction**. The greater n, the less thick a lens is needed to do the same job. Air is almost like the vacuum: $n = 1.00029$. Polyurethane lenses used in eyeglasses have an index of refraction as high as $n = 1.66$. The glass used in lenses has an index of 1.52. Diamond's $n = 2.4$; such lenses would be scratchproof and very thin, but not cheap.

[17]We refer the reader to the beautiful book *QED* by Richard Feynman [Fe], where this remarkable puzzle is explained.

Figure 3.20. Light travels slower in glass; hence the ray avoids the thicker part of the lens.

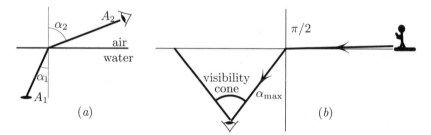

Figure 3.21. (*a*) Snell's law; (*b*) for the diver, the world above the water is "compressed" into a cone.

A (very) brief history. This minimality property of light was observed by Heron of Alexandria (c. AD 60), who stated that the ray traveling in the air between two points and reflecting off flat mirrors between the points is shorter than any nearby path. The general form of the principle, applicable to general media, was stated by Fermat in 1662. Perhaps understandably, some people objected: the principle, they said, ascribes foreknowledge to nature. The skeptics should not have worried. Fermat's principle can be explained via classical electrodynamics, by appealing to the wave nature of light. The current explanation, outlined in Feynman's book mentioned earlier, is provided by quantum electrodynamics, where the wave is replaced by the probability wave function.

Fermat's principle determines how the rays bend in crossing a boundary between two media. A more explicit consequence of this principle is known as Snell's law.

Snell's law of refraction.[18] *For a ray of light crossing the interface between two media (see figure 3.21), the sines of the respective angles α_1, α_2 between the ray and the normal to the interface are proportional to the speeds c_1 and c_2 of light in the respective media:*

$$\frac{\sin \alpha_1}{c_1} = \frac{\sin \alpha_2}{c_2}. \tag{3.12}$$

In other words: the quantity $\frac{\sin \alpha}{c}$ remains unchanged in crossing the media interface.

This law is restated as the "lifeguard problem" in section 3.13.

We literally see the world through Snell's law: each ray hitting a "pixel" on our retina gets there while obeying (3.12).

Problem. *What is the physical meaning, if any, of $\frac{\sin \alpha}{c}$?*

Solution. $\sin \alpha/c$ is the speed of the point where the wavefront intersects the media interface (figure 3.22). The explanation is provided by figure 3.23, as follows. The figure shows the wavefront in the air at two instants one second apart. Since the rays are perpendicular to the wavefront,[19] $\triangle PP'Q$ is a right triangle, where the meaning of Q is explained by the figure. Since $QP' = c_1$, we have $PP' = c_1/\sin\alpha_1$. This is the displacement of P in one second and therefore is the speed in question.   ◇

The critical angle. Figure 3.21 explains an interesting phenomenon observed by divers and snorkelers. As you look at the surface of

---

[18]See problem 3.13 for a mechanical proof of the fact that Snell's law follows from Fermat's principle.

[19]This intuitively very plausible statement can be justified via Huygens's principle; the details can be found in [ARN]. The fact depends on the medium being isotropic. For anisotropic media the rays need not be perpendicular to the front, at least not in the sense of Euclidean orthogonality.

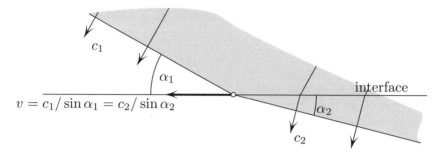

Figure 3.22. The break in the front has speed $v = c_1 / \sin \alpha_1 = c_2 / \sin \alpha_2$.

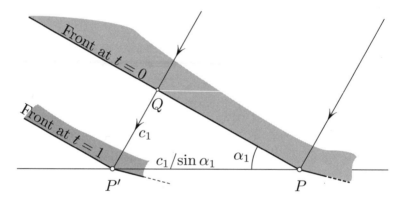

Figure 3.23. Computing the speed of the break point in the front.

the water from below, everything above water appears to be confined to a cone. Outside a circle on the surface of the water, with center above the eye of the observer, transmitted light does not reach the eye; only light reflected from below reaches the eye. Here is why. A nearly horizontal ray in the air forms an angle $\alpha_{air} = \pi/2$ with the vertical before hitting the water; once in the water, the angle with the vertical is $\alpha_{water} = \sin^{-1} \frac{c_{water}}{c_{air}}$, according to Snell's law. For divers looking up, this is the maximal angle with the vertical at which they see the world above.

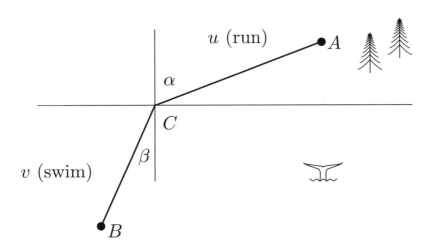

Figure 3.24. How to place C to get from A to B in shortest time?

3.13 Saving a Drowning Victim by Fermat's Principle

Problem. *A lifeguard at A wants to save a swimmer at B. The lifeguard runs with speed u and swims with speed v. What point C in figure 3.24 gives shortest possible time?* Assume that the swimmer's speed is zero.[20]

This problem is identical to the preceding problem of finding the time-minimizing path of the refracting ray of light. Just like the ray of light, our lifeguard follows Fermat's principle of least time, which implies that the angles in figure 3.24 satisfy Snell's law. The problem, in other words, is asking to explain how Snell's law follows from Fermat's principle.

The standard textbook proof is to write the time as a function of the unknown position x of C, and to find the minimizing x by differentiation. The solution given here involves no calculation and no calculus.

---

[20] Including, especially, his speed in the vertical direction.

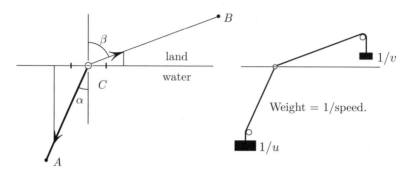

Figure 3.25. Tensions can be created either by constant tension springs or by weights.

Solution. Figure 3.25 shows a ring sliding along the line (the "shoreline"); the ring is connected to the point A by a constant tension spring[21] whose tension we choose to be $1/u$, the reciprocal of the running speed. Similarly, we connect the ring to the point B by another spring of constant tension $1/v$, the reciprocal of the swimming speed.

Since the potential energy of a constant tension spring equals its length times its tension, the potential energy of our mechanical system is

$$AC/u + CB/v. \tag{3.13}$$

But this expression coincides with the time of travel from A to B! We thus endowed the time with the mechanical meaning of the potential energy.

If the travel time is minimal, then the energy is minimal and hence the system is in equilibrium; thus, in particular, the forces on the ring along the line are in balance:

$$\frac{1}{u}\sin\alpha = \frac{1}{v}\sin\beta.$$

We have thus reproduced Snell's law. Along the way we also discovered a mechanical interpretation of Snell's law: the forces on the the ring in the direction of the interface are in balance.

[21] Such springs are described in section A.1.

The lesson for the lifeguard is to run so that the angles α and β with the normal to the shoreline satisfy Snell's relation. $\diamond$

Here is an amusing interpretation of the solution to the lifeguard problem (this interpretation is a restatement of the observation in the problem on page 55). Imagine the guard carrying a long pole always held perpendicular to the direction of his motion, whether he runs or swims. This pole is the analog of the wavefront. (Alternatively we can think of a row of people running/swimming side by side, all having exactly the same athletic ability.) Consider the point of intersection of the pole with the shoreline. This point moves with constant speed while the lifeguard runs and, generally speaking, with another speed as he swims. Now the best strategy is distinguished by the property that the two speeds are the same.

3.14 The Least Sum of Squares to a Point

Problem. *Given three points A, B, and C in the plane, find the point M for which the sum of* **squares**

$$AM^2 + BM^2 + CM^2 \qquad (3.14)$$

is minimal.

Solution. The sum of squares (3.14) can be interpreted as the potential energy of three zero-length springs[22] connecting M to the points A, B, and C, with Hooke's constant $k = 2$ for each spring (see figure 3.26). Now if this energy (3.14) is minimal, then the sum of forces upon M is zero:

$$\overline{MA} + \overline{MB} + \overline{MC} = \mathbf{0},$$

where we divided by the common factor $k = 2$. This amounts to the statement that M *is the center of mass of A, B and C.* Thus M is

---

[22]For background, see section A.1. We need only two facts here: first, that the tension of a zero-length spring is $k \cdot L$, where L is the length of the spring and where k is Hooke's constant, and second, the potential energy of the spring is $\frac{1}{2}kL^2$.

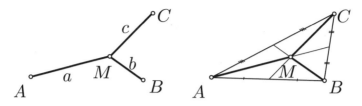

Figure 3.26. Center of mass of three points minimizes (3.14).

the intersection of the medians of $\triangle ABC$ (for an explanation see the next problem).   ◇

An alternative solution. We observe that (3.14) is the moment of inertia[23] with respect to M of three masses $m_1 = m_2 = m_3 = 1$ located at A, B, C. But the moment of inertia is minimal when taken with respect to the center of mass (see discussion in section A.9). ◇

3.15 Why Does a Triangle Balance on the Point of Intersection of the Medians?

Problem. *Why does the center of mass of a triangular piece of a cardboard coincide with the point of intersection of its medians?*

Solution. The key is to explain why the triangle balances on each of its medians, say AA', when placed on a horizontal knife edge as shown in figure 3.27. Imagine slicing our triangle into thin strips parallel to the side BC; one such strip is shown in figure 3.27. *The median AA' bisects each segment of the strip parallel to the side BC,* and therefore each strip will balance on the knife edge[24]. Hence the triangle balances on AA'. Now the point of balance (M) must lie on the line of balance, that is, on AA'. Since this applies to all medians, the point of balance lies on their intersection.   ◇

---

[23] The concept, along with everything used here, is explained in section A.9.
[24] We are ignoring a possible error near the ends of the strips, but the relative size of the error approaches zero with the thickness of the strip.

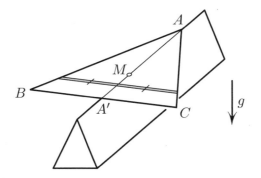

Figure 3.27. The triangle balances on a median since each thin strip is bisected by the median.

Remark. *Concentrating the mass of the cardboard triangle in the vertices in equal amounts per vertex will leave the center of mass unchanged. Indeed, the triangle with the redistributed mass still balances on AA' since B balances C (A' is the midpoint of BC) while A lies on AA' and thus does not affect the balance of the other two masses.*

3.16 The Least Sum of Distances to Four Points in Space

Problem. *Given four points A, B, C, and D in space, consider the sum of distances $AX + BX + CX + DX$ to a point X. What point X minimizes this sum?*

Solution. Using mechanics we will show that

$$\angle AXB = \angle CXD$$

and that, moreover, *these two angles are bisected by the same line.* Since the points A, B, C, and D have equal rights, the same result will then hold for any different pairing of these points. For instance, we will then know that $\angle AXC = \angle BXD$ and these angles share a common bisector.

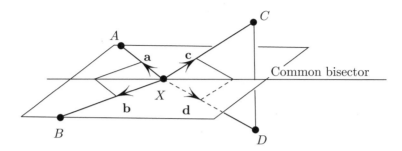

Figure 3.28. If X minimizes $AX + BX + CX + DX$, then the marked angles share a bisector and are equal.

Here is the proof. Let us take four identical constant tension springs[25] of tension $T = 1$ and connect them as shown in figure 3.28. The length of each spring is equal to its potential energy, and the combined length is equal the potential energy of the system. Hence the minimal-length configuration corresponds to an equilibrium, and hence to the vanishing of the sum of forces $\mathbf{a}$, $\mathbf{b}$, $\mathbf{c}$, and $\mathbf{d}$ acting upon the common point X:

$$\mathbf{a} + \mathbf{b} = -(\mathbf{c} + \mathbf{d}). \qquad (3.15)$$

Our springs have identical constant tensions: $|\mathbf{a}| = |\mathbf{b}|$; hence the vector $\mathbf{a} + \mathbf{b}$ lies on the bisector of the angle AXB; similarly, $\mathbf{c} + \mathbf{d}$ lies along the bisector of the angle CXD. By (3.15), these bisectors lie on the same line; we proved that the bisector is shared by the two angles.

To show that the angles AXB and CXD are equal, we note that $|\mathbf{a}| = |\mathbf{b}| = 1$ implies $|\mathbf{a} + \mathbf{b}| = 2\cos\angle AXB$. Similarly, $|\mathbf{c} + \mathbf{d}| = 2\cos\angle CXD$. From (3.15), we conclude that $\angle AXB = \angle CXD$. $\diamond$

---

[25] Such springs are described in section A.1.

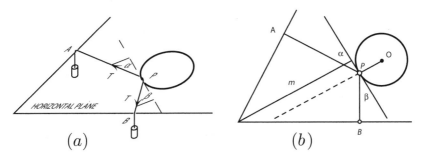

Figure 3.29. If the sum $PA + PB$ is minimal when $\alpha = \beta$.

3.17 Shortest Distance to the Sides of an Angle

Problem. *A circle lies inside an acute angle. From which point on the circle is the sum of distances to the sides of the angle minimal?*

Solution. Figure 3.29 shows the angle and the circular hoop lying in the horizontal plane. A string is looped through the hoop, with the ends of the string thrown over the angle's sides and holding two equal weights. The string is perfectly frictionless, so that it is perpendicular to the angle's sides at A and B. Thus PA and PB measure the distances from P to the angle's sides.

The length $AP + PB$ is proportional to the potential energy of our mechanical system—indeed, the larger $AP + PB$, the larger the sum of elevations of the two weights. Thus the minimum of $AP + PB$ corresponds to an equilibrium, and thus to the balancing of the tensions at P: $T \cos \alpha = T \cos \beta$, that is,

$$\alpha = \beta.$$

This relation characterizes the "best" point P and solves the problem.   ◇

An explicit answer. The condition $\alpha = \beta$ says that the tangent line at P is perpendicular to the angle's bisector m, as seen from figure 3.29(*b*). This, in turn, amounts to OP being parallel to m. This is more explicit: *the "best" P is the tip of the radius that is parallel*

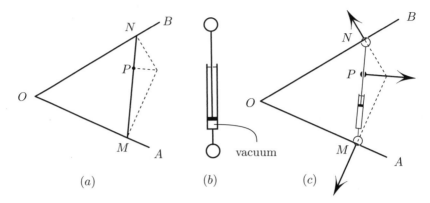

Figure 3.30. For the shortest segment through P the three dashed perpendiculars meet at one point.

to the angle's bisector and that points toward the vertex of the angle. The "worst" point—the one maximizing $AP + PB$—is on the radius parallel to m pointing outward.

3.18 The Shortest Segment through a Point

Problem. *Let MN be the shortest of all segments lying inside a given angle AOB (with M lying on OA and N lying on OB) and passing through a given point P inside the angle. Prove that the three perpendiculars at the points P, M, and N, as shown in figure 3.30, are concurrent.*

Solution. Consider a device in figure 3.30(*b*)—essentially a rod which tries to shorten, with a ring at each end. We install the device as shown in part (*c*): the rod slides freely through a sleeve at P which can rotate, and the rings can slide without friction along the angle's sides.

   If the segment *MN* has minimal length, then the mechanical system has the minimal potential energy and then the rod with its rings is in equilibrium. Thus all three normal reaction forces shown in figure 3.30 acting upon rod and rings add up to zero, and so

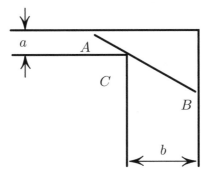

Figure 3.31. What is the largest length of a pole that can be carried around the corner?

do their torques. The lemma on concurrent forces (section A.6) states that if the sum of three forces is zero and the sum of the torques of these forces is zero, then the lines of the three forces are concurrent.   ◇

3.19 Maneuvering a Ladder

Problem. *Two hallways of respective widths a and b meet at a right angle. What is the largest length pole that can be passed around the corner from one hallway to the other?*

Solution. The problem is to find the segment ACB of minimal length, passing through the corner C, with the ends lying on the outer walls. For example, imagine carrying a telescoping ladder around the corner and shortening it if it doesn't fit; once the ladder clears the corner, its length will be the minimal length mentioned above.

A mechanical "analog computer" shown in figure 3.32 solves the problem. A telescoping rod (a vacuum-filled cylinder with a piston, as shown in figure 3.30) has two rings (A and B) welded at the two ends. The rings are slipped over the lines MN and NP as shown. The rod can slide through a freely pivoting sleeve at C.

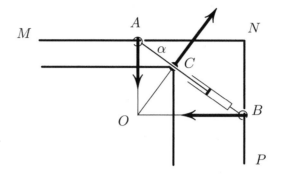

Figure 3.32. A solution using two rings A and B, a rod pivoting on C, and a spring.

The potential energy of our mechanical system is a monotone function of the length AB, and thus the position of minimal length corresponds to the equilibrium. Now the rod/piston/rings system is subject to three forces shown in figure 3.32. In equilibrium, the sum of forces and of their torques of the three forces is zero, and hence the lines of the forces are concurrent.[26] This solves the problem: *for the minimal length segment the three normals[27] to the walls at A and B and to the line AB at C are concurrent.*   ◇

Our geometric solution yields an analytic expression for the angle $\alpha = \angle NAC$ at once. Indeed, from $\triangle OCA$ and $\triangle OCB$ we have

$$OC = AC \cot \alpha = CB \tan \alpha.$$

Substituting $AC = a/\sin \alpha$ and $CB = b/\cos \alpha$ into the previous expression and simplifying yields

$$\tan \alpha = \left(\frac{a}{b} \right)^{\frac{1}{3}}.$$

Again, this solution is much quicker than the traditional calculus solution.

---

[26] See the lemma on concurrent forces, section A.6.

[27] Note that the forces are indeed perpendicular to the corresponding lines since there is no friction.

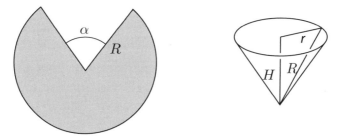

Figure 3.33. What angle α will maximize the volume of a cup?

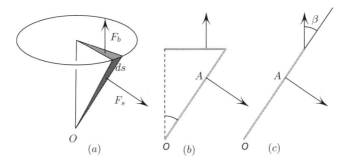

Figure 3.34. The sum of torques upon the shaded sector is zero.

3.20 The Most Capacious Paper Cup

Problem. *I want to make a conical paper cup from a sector of a paper disk, by gluing together the edges of the cut shown in figure 3.33. The radius R of the disk is fixed. For the cone of largest possible volume, what is the angle β between the cone's generator and the axis?*

Solution. The problem is to find the shape of the cone of largest possible volume, from among all cones with the given length of the generator.

The mechanical "analog computer" consists of a bouquet of segments of length R—these will form the generators of the cone; all generators are tied together at one point O fixed in space; the other ends of these generators are constrained to a plane (the base of the cone) by a *frictionless constraint*.[28] The plane is free to change its distance to O. The resulting cone is filled with compressed gas. The plane, no longer free, is pushed away from O by gas and pulled toward O by the segments. The volume of the cone is a decreasing function of the potential energy. Hence the maximal volume corresponds to the minimal potential energy and thus to an equilibrium. In particular, an infinitesimal sliver of generators (figure 3.34) will be in equilibrium. The torque of the forces upon the sliver around O is therefore zero, which solves the problem; it remains only to decipher this zero-torque statement. We will not need to know the forces upon the sliver at O; there are exactly two remaining forces acting on this sliver, as seen in figure 3.34(*b*): (i) the force of outward pressure (the subscript *s* stands for the "side")

$$F_s = pA_s, \tag{3.16}$$

where A_s is the area of the sliver, and (ii) the force of constraint from the plane (the subscript *b* stands for the base)

$$F_b = pA_b, \tag{3.17}$$

where A_b is the area of the sector in the base. Note that the generators do not interact with each other directly.

Now, the opposing torques of these two forces relative to O are in balance, as in figure 3.34(*c*):

$$F_s \cdot OA = (F_b \sin \beta) \cdot R.$$

Substituting (3.16) and (3.17), we get

$$A_s \cdot OA = A_b \sin \beta \cdot R. \tag{3.18}$$

But $OA = \frac{2}{3}R$, since the centroid of any triangle is two-thirds of the way from the vertex to the base. In addition, $A_b/A_s = r/R = \sin \beta$ since the ratio of the areas of the two triangles with a common base (ds) equals the ratio of their heights. Substituting this into (3.18) we

---

[28] One can imagine the tips of the generators to be magnetically stuck to the plane but able to slide without friction.

obtain

$$\frac{2}{3} = \sin^2 \beta,$$

which solves the problem.   ◇

3.21 Minimal-Perimeter Triangles

A mechanical argument suggested this theorem, only a few years old.[29]

Theorem. *Let K be an arbitrary closed convex planar curve containing no straight segments. Let $\triangle ABC$ have minimal[30] perimeter among all triangles containing K. Then:*

1. *The three segments connecting a vertex of $\triangle ABC$ with the tangency point on the opposite side of $\triangle ABC$ are concurrent, that is, they meet at one point; equivalently, by Ceva's theorem[31]*

$$abc = a'b'c',$$

   where a, a', b, b', c, c' are the lengths shown in figure 3.35(a).
2. *The three perpendiculars to the sides of $\triangle ABC$ at the tangency points are concurrent (figure 3.35(b)).*

Proof. *A proof by mechanics* goes as follows.

A mechanical system. Consider three (infinite) rods forming a triangle ABC, with each pair of rods slipped through a small ring, as shown in figure 3.36. The rods are in frictionless contact with the rings, and thus can form a triangle of any shape, except for the constraint we impose: $\triangle ABC$ must contain the curve K in its interior, that is, K is an obstacle impenetrable to the rods. Now let us connect each pair of rings by a constant-tension spring as shown

---

[29] For a rigorous proof, see [L1]).
[30] "Minimal" can be replaced by "critical."
[31] For the statement and a proof by mechanics see problem 5.6.

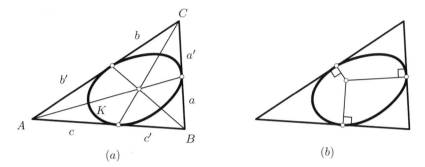

Figure 3.35. Minimality of the perimeter of $\triangle ABC$ implies the concurrency of (a) the Cevians and (b) the normals.

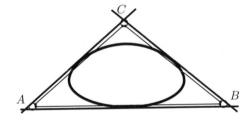

Figure 3.36. The rods-rings-springs mechanical computer.

in the figure. The springs are trying to collapse $\triangle ABC$, but the obstacle K prevents such a collapse. Recall that our springs have tension $T = 1$, so that the potential energy of a spring equals its length;[32] we thus endowed the perimeter with the physical meaning of potential energy. If $\triangle ABC$ has minimal perimeter, then the mechanical system is in equilibrium.

CONCURRENCY OF THE PERPENDICULARS (FIGURE 3.35(b)) IS NOW IMMEDIATE. The assembly of the three rods, rings, and springs, considered as one system, is subject to precisely three reaction forces $\mathbf{F}_k$ from the obstacle K. In equilibrium, the sum of these forces vanishes, as does the sum of their torques. But the lemma

---

[32]For a one-line explanation see section A.1.

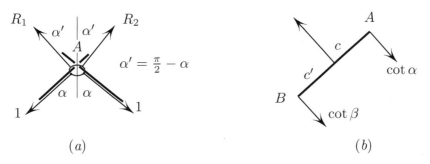

Figure 3.37. Force balance on the ring (a) and torque balance on the rod (b).

on concurrent forces (section A.6) states that in that case the lines of the three forces are concurrent. This proves the second part of the theorem.

PROOF OF $abc = a'b'c'$. For a minimal perimeter $\triangle ABC$ each rod is subject to the three normal reaction forces shown in figure 3.37: one force from K and one from each of the two rings that are in contact with the rod. For each rod, the sum of the torques of the three forces relative to the point of tangency is zero; this will lead to (3.19)–(3.20) below, as we will show momentarily; multiplying these equations together will give $abc = a'b'c'$. All we need to do is find the reaction force from each ring upon the rod. To be specific, let us pick the ring A and the rod AB. The ring is subject to four forces: two from the springs and two from the rods, as in figure 3.37(a).

Projecting these four forces first onto the bisector of angle A and then onto the perpendicular to the bisector we obtain from the force balance

$$R_1 \cos \alpha' + R_2 \cos \alpha' = 2 \cos \alpha, \quad R_1 \sin \alpha' - R_2 \sin \alpha' = 0,$$

where $\alpha' = \frac{\pi}{2} - \alpha$. From the second equation $R_1 = R_2$ (an interesting fact in itself!), and from the first, using $\cos \alpha' = \sin \alpha$ we obtain

$$R_1 = \cot \alpha.$$

This is the reaction force from the rod AB upon the ring A. By Newton's third law, the rod AB feels the same force back from the

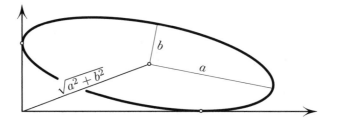

Figure 3.38. The distance from the center of the ellipse to the origin is constant.

ring. We recall that the force is normal to the rod, since the ring is frictionless. Figure 3.37(b) summarizes all the forces acting on the rod AB. The torque balance of these forces relative to the tangency point gives

$$c \cot \alpha = c' \cot \beta. \tag{3.19}$$

Similarly, we have

$$a \cot \beta = a' \cot \gamma,$$
$$b \cot \gamma = b' \cot \alpha. \tag{3.20}$$

Multiplying the three equations together, as mentioned earlier, and canceling, we obtain $abc = a'b'c'$. This relation, in turn, implies the concurrency of the Cevians by Ceva's theorem.[33]   ◇

3.22 An Ellipse in the Corner

This problem comes from a Putnam competition.

Problem. *Consider an ellipse lying in the first quadrant of the (x, y) plane and tangent to the coordinate axes. Prove that the distance from the center of the ellipse to the origin depends only on the semiaxes of the ellipse and not on its orientation.*

---

[33] We prove this theorem using centers of mass in section 5.6. A very nice geometrical treatment of Ceva's theorem can be found in [CG].

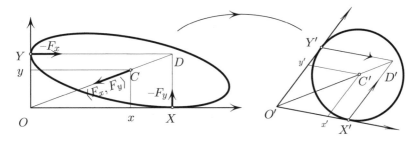

Figure 3.39. The ellipse is in neutral equilibrium if the two rectangles are similar.

As a consequence of this result, if we rotate the ellipse while keeping it tangent to the axes, its center would trace an arc of a circle. In the easier degenerate case when the ellipse is a line segment, the arc is a quarter circle.

Solution. Here is a solution of this problem using mechanics.

*A **mechanical system**.* Think of the ellipse as a rigid object sliding without friction along the coordinate axes. A stretched spring ties the center of the ellipse to the origin (figure 3.39). The potential energy of the spring is an increasing function of its length,[34] and hence *minimality of the length is equivalent to minimality of the potential energy, that is, to the equilibrium.* To summarize: it suffices to prove that the ellipse is in equilibrium in any orientation.

AN EQUILIBRIUM CONDITION. To prove that the ellipse is in equilibrium in any position, it suffices to show that the two rectangles shown in figure 3.39 are similar.

Indeed, three forces act upon the ellipse (two reactions and one spring), and for the sum of torques of these forces to vanish it is necessary and sufficient that the lines of the forces pass through a single point, according to the lemma on concurrent forces in section A.6. The latter condition is equivalent to the similarity of the two rectangles. Our problem has been reduced to proving the

---

[34]The particular nature of the spring does not matter in this solution.

similarity of the two rectangles, an interesting property in its own right, but how to prove it?

FINISHING THE PROOF. I do not want to assume that the reader is familiar with linear transformations, so this discussion does not use them. Instead, imagine drawing figure 3.39 on a sheet of Plexiglas and tilting the sheet under the sun so that the shadow of the ellipse on the flat ground is a circle. Since parallel lines cast parallel shadows, our two rectangles become parallelograms. In fact, these parallelograms are rhombi, as we will show shortly, and since these rhombi share an angle, they are similar to each other. The original rectangles are then similar as well, since projection preserves similarity. It remains to explain why the rectangles turn into rhombi. The parallelogram $O'x'C'y'$ is a rhombus since its diagonal $O'C'$ bisects the angle, which follows from the facts that C' is the center of the circle and that OX' and $O'Y'$ are tangent to the circle. The parallelogram $O'X'D'Y'$ is a rhombus since $O'X' = O'Y'$, as the two tangent segments to a circle.   ◇

3.23 Problems

The challenge is to find physical solutions to the following problems.

1. Of all the rectangles with the given area S find the one with the largest possible perimeter.
2. Of all rectangular solids of given volume, find the one with least surface area.
3. Find the right triangle of maximal area with the given sum of the lengths of one leg and a hypotenuse.
4. Find the rectangle of largest area given that two vertices of the rectangle lie on a chord of a given circle, and the other two vertices lie on the circle, on the smaller of the two arcs.
5. Inscribe the rectangle of largest perimeter into a triangle of base b and the altitude h.
6. Find the dimensions b and h of the rectangle inscribed in a circle with the maximal value of bh^2. (Note: This is the problem of

cutting the stiffest joist out of a given log; bending stiffness of a joist is proportional to bh^2.)

7. Inscribe a rectangular solid of largest volume into the hemisphere of a given radius.

8. Inscribe the cylinder of largest volume into a sphere of the given radius.

9. Circumscribe a cone of least volume around a sphere of the given radius R.

10. Find the cylinder of greatest surface area inscribed in the cone whose axial section has angle 2α at the vertex and whose base has radius R.

11. Find the tangent line to the ellipse $x^2/a^2 + y^2/b^2 = 1$ which cuts the triangle of least area off of the first quadrant.

12. A solid consists of a cylinder with a hemisphere on top. What proportions of this solid will minimize its area for a given volume?

13. Find the relative sizes of a sphere and a cube with a given combined volume and of most surface area.

14. A transversal section of a channel is in the form of an isosceles trapezoid. What slope of the sides will minimize the "wet perimeter" of the section, given the area S of the trapezoid and the depth h?

15. A railroad track passes through a warehouse B, and lies at a distance a from a town A. A straight road must be built from the track to the town. What angle should this road form with the track to minimize the cost of deliveries, given that the cost of transportation along the road is n times more than along the railroad $(n > 1)$?

16. Consider a triangle of maximal perimeter inscribed in a convex curve K. Show that the two adjacent sides form equal angles with the tangent to K at the common vertex. Show that the same property holds for any inscribed n-gon of maximal (or even of critical) perimeter.

17. (An open problem.) Consider a tetrahedron of least area circumscribed around a given convex body K. Give a geometrical characterization of such a tetrahedron.

4

INEQUALITIES BY ELECTRIC SHORTING

4.1 Introduction

The following short outline of the necessary background should suffice for the reading of this chapter. More on the concepts described below is in the appendix.

Electrical current. The current in a copper wire is the flow of the "gas" of electrons in the ionic lattice of copper, analogous to the flow of water in a pipe. Just like the flux of water through the pipe is measured in gallons per second, the electric current is measured in units of charge per second, passing through a cross section of the wire. The current is denoted by I and is expressed in coulombs per second, or amperes.

Voltage. Let us consider a steady flow of water in a long pipe. Because of the friction with walls, the water wants to slow down. Since the flow is assumed steady, the pressure is higher upstream; it is this pressure gradient that pumps the water at a steady rate. In the same way, a steady current through a wire requires constant application of electrical pressure, called the voltage. The difference in voltage propels the electrons along the wire.

Resistance. To simplify things, imagine water flowing steadily though a pipe with no friction, but now the pipe has a porous obstruction. The pressure difference across the obstruction keeps the flow steady. Note also that the more water we pump per second, the greater is the pressure difference. Similarly, consider a wire

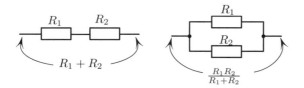

Figure 4.1. Resistors in series add; conductivities in parallel add.

with a resistor such as a filament of a light bulb—an analog of an imperfect clog in a pipe. Assume a steady current I through the wire. In complete analogy with water, there is a voltage difference V across the resistor, and, just as with water, the greater the current, the greater is this voltage. In fact, experiments show a linear relationship $V = RI$ (Ohm's law), where R is a constant. This constant is called the resistance—a very reasonable name, since a large R signifies a large V ('pressure difference') for a given I (flux).

Resistors in series and in parallel. For two resistors R_1 and R_2 connected in series (see figure 4.1), the resistance of the resulting combination is the sum:

$$R = R_1 + R_2. \tag{4.1}$$

For two resistors connected in parallel, the resulting resistance R satisfies

$$\frac{1}{R} = \frac{1}{R_1} + \frac{1}{R_2}. \tag{4.2}$$

A short proof of this is given in the appendix; here I only mention that both rules make intuitive sense. The second of these perhaps requires an explanation. When two resistors are connected in parallel, the resulting circuit conducts better, since it creates two channels for the current. One could guess that the conductivity is the sum of conductivities of each channel. In fact, this is precisely what the rule (4.2) states, if we define the conductivity to be the reciprocal of the resistance.

Kirchhoff's second law. The sum of currents entering a node where several wires join equals the sum of currents exiting the node.

Figure 4.2. Kirchhoff's second law.

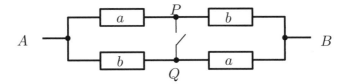

Figure 4.3. Shorting P with Q decreases resistance, explaining (4.3).

For the example in figure 4.2, the law gives $I_1 + I_2 + I_3 = I_4 + I_5$. This law expresses the fact that the charge cannot accumulate in a node, just as water cannot accumulate at a junction of several pipes.

4.2 The Arithmetic Mean Is Greater than the Geometric Mean by Throwing a Switch

The circuit shown in figure 4.3 is made of resistances a and b. Starting with the switch open as shown, each of the two parallel paths has resistance $a + b$. Since these paths are in parallel, the total resistance between A and B is $\frac{a+b}{2}$, according to (4.2). Let us now close the switch. The resistance of the circuit with a short is the same as or less than before.[1]

What is this new, smaller resistance? We now have two resistors in sequence, each of strength $(a^{-1} + b^{-1})^{-1} = \frac{ab}{a+b}$. Thus

$$\frac{a+b}{2} \geq \frac{2ab}{a+b}. \tag{4.3}$$

[1] This intuitively obvious statement should be proven rigorously. We will give such a proof to satisfy skeptics, but our focus here is on establishing connections (algebra and circuits in the present problem) rather than a proof.

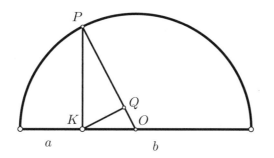

Figure 4.4. Arithmetic, geometric, and harmonic means: $PO = (a + b)/2$, $PK = \sqrt{ab}$, $PQ = ((a^{-1} + b^{-1})/2)^{-1}$.

This implies the inequality between the arithmetic and the geometric means stated in the title of the section.

Remark 1. Equation (4.3) implies not only the inequality between the arithmetic and geometric, but also with the harmonic mean $((a^{-1} + b^{-1})/2)^{-1} = 2ab/(a + b)$:

$$\frac{a + b}{2} \geq \sqrt{ab} \geq \frac{2ab}{a + b}. \tag{4.4}$$

Indeed, if $A \geq B > 0$, then $\sqrt{A} \geq \sqrt{B}$; multiplying this inequality first by $\sqrt{A}$ and then by $\sqrt{B}$ we obtain $A \geq \sqrt{AB} \geq B$. Now treating the two sides in (4.3) as A and B and applying the last inequality gives (4.4).

Remark 2. (A geometrical interpretation of the inequality (4.4).) Let a and b be the lengths of two abutting segments, and consider a semicircle whose diameter is the union of these two segments. Construct the perpendicular from the point K where the two segments abut, and let P be a point of intersection of this perpendicular with the semicircle. Let O be the center of the semicircle and let Q be the foot of the perpendicular from K onto the radius OP. One can show that

$$PK = \sqrt{ab}, \quad PO = \frac{a + b}{2}, \quad \text{and} \quad PQ = \frac{2ab}{a + b}.$$

This gives a geometrical interpretation/proof of the inequality (4.4).

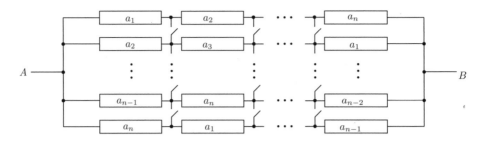

Figure 4.5. Closing all switches decreases the resistance between A and B.

4.3 Arithmetic Mean $\geq$ Harmonic Mean for n Numbers

Recall the definition of the harmonic mean: each number is inverted; the arithmetic mean of these inverses is found and then the result is inverted again. Loosely speaking, the harmonic mean is the arithmetic mean through the lens of inversion.

Here is an electric proof of the fact that the arithmetic mean is the greater of the two means:

$$\frac{1}{n}\sum_{k=1}^{n}a_k \geq \left(\frac{1}{n}\sum_{k=1}^{n}a_k^{-1}\right)^{-1}. \tag{4.5}$$

Each row in figure 4.3 consists of the same resistances. Note that each subsequent row is a cyclic permutation of the one before; thanks to this fact, each column also consists of the same resistances.

We start with all switches open; by throwing them all at the same time we will prove the above inequality. Here are the details.

All switches open. All rows have the same resistance $\sum_{k=1}^{n}a_k$; the n identical resistances in parallel give the effective resistance between A and B:

$$\frac{1}{n}\sum_{k=1}^{n}a_k.$$

All switches closed. Each column consists of n parallel resistors, and thus has resistance $(\sum_{k=1}^{n}a_k^{-1})^{-1}$, as proven in the appendix in

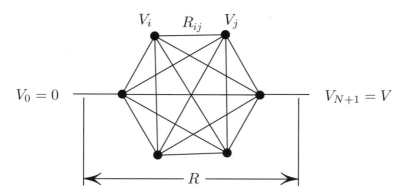

Figure 4.6. The voltages on vertices minimize the dissipated power (4.6).

section A.14. With n such columns in series, the resistance between A and B is n times greater:

$$n \left(\sum_{k=1}^{n} a_k^{-1} \right)^{-1} = \left(\frac{1}{n} \sum_{k=1}^{n} a_k^{-1} \right)^{-1}.$$

This proves the claim (4.5).   ◇

4.4 Does Any Short Decrease Resistance?

We are considering a network of resistors; an example is shown in figure 4.6. Such a collection of dots connected by lines is called a *graph*. The dots are called the vertices, and the lines are called the edges of the graph. In our picture the edges are the resistors connecting the vertices. Any two vertices i and j are connected by a resistor R_{ij}. We allow $R_{ij} = \infty$, meaning that not all pairs of vertices need be electrically connected. The value $R_{ij} = 0$ corresponds to short between the vertices i and j.

Rayleigh's monotonicity law. Let us fix two vertices in the graph and consider the effective resistance R between them, as seen in figure 4.6; R is a function of R_{ij}. It is intuitively very plausible that *if any R_{ij} increases, then the overall resistance R increases as well.* This statement is known as *Rayleigh's monotonicity law.*[2] In the preceding two sections we used a special case of this principle: by creating a short, we changed some resistances from ∞ to 0, thus decreasing the overall resistance.

Here is the proof of Rayleigh's monotonicity law. To fix notations, let our graph consist of $N + 2$ vertices $v_0, v_1, \ldots, v_N, v_{N+1}$ (figure 4.6), and let us consider the resistance R between v_0 and v_{N+1}. To fix things, we impose a voltage V between v_0 and v_{N+1}. We can choose $V_0 = 0$ (i.e., we can ground the vertex v_0), and thus we have $V_{N+1} = V$.

If V is the voltage across a resistor R, then the power dissipated on the resistor is V^2/R (this is explained in a short paragraph in the appendix in section A.16.) Thus **if** we somehow maintained voltages $\tilde{V}_k$ on the vertices v_k, then the power dissipated in the network would be

$$P(\tilde{V}_1, \ldots, \tilde{V}_N) = \sum_{i,j=0}^{N+1} \frac{(\tilde{V}_i - \tilde{V}_j)^2}{R_{ij}}, \qquad (4.6)$$

where $\tilde{V}_0 = V_0, \tilde{V}_{N+1} = V_{N+1} = V$.

In reality, we are controlling only the voltage $V_{N+1} = V$, letting the other "free" voltages $V_k, 1 \le k \le N$ "decide for themselves" what to be.

Here is a beautiful fact: *The actual power P dissipated on the network is least possible with the given imposed voltage V* (4.6):

$$P = \min_{\tilde{V}_1,\ldots,\tilde{V}_N} \sum_{i,j=0}^{N+1} \frac{(\tilde{V}_i - \tilde{V}_j)^2}{R_{ij}}, \quad V_0 = 0, \ V_{N+1} = V. \qquad (4.7)$$

I explain this in the next paragraph, but let us first use this fact to complete the proof of Rayleigh's principle. Since R_{ij} are in the denominators, P is a *decreasing* function of each R_{ij}. But since $P = V^2/R$, that is, $R = V^2/P$, we conclude that R is an *increasing* function of each R_{ij}. This proves Rayleigh's principle, except that we have to verify (4.7).

---

[2] We refer to a very nice book [DS] for further details and references.

Let $(V_1, \ldots, V_N)$ be the actual voltages on the vertices $v_1, \ldots, v_N$. We have

$$\frac{1}{2}\frac{\partial}{\partial V_k}P(V_1, \ldots, V_N) = \sum_{i=0}^{N+1} \frac{V_k - V_i}{R_{ki}} \overset{\text{Ohm}}{=} \sum_{i=0}^{N+1} I_{ki} \overset{\text{Kirchhoff}}{=} 0.$$

(The background on Ohm's and Kirchhoff's laws is presented in sections A.12 and A.13.) But $P(\tilde{V}_1, \ldots, \tilde{V}_N)$ is a positive quadratic function, and thus has only one critical point which is a minimum. We showed that *the actual voltages V_k minimize the power function.* The power dissipated on the network is indeed given by the minimum of the power function, which proves (4.7).   ◇

4.5 Problems

1. Prove that the inequality

$$\frac{1}{\frac{1}{a+b} + \frac{1}{c+d}} \geq \frac{1}{\frac{1}{a} + \frac{1}{c}} + \frac{1}{\frac{1}{b} + \frac{1}{d}}$$

   holds for any positive numbers a, b, c, d. Hint: Consider the circuit in figure 4.3, and change some resistances. (For more references on this approach, see [DS].)

2. Give a mechanical interpretation of the expression (4.6) involving springs.

3. Find a mechanical analog of the voltage, the current, the resistance, Kirkhoff's second law, and Ohm's law. Hint: Use springs with Hooke's constants $k_{ij} = 1/R_{ij}$.

4. Give a mechanical interpretation of power-minimization principle.

5. Find a mechanical analog of (4.1) and (4.2) involving springs.

5

CENTER OF MASS: PROOFS AND SOLUTIONS

5.1 Introduction

The concept of the center of mass was used by Archimedes more than 2,400 years ago.[1] Much later Euler introduced another mass-related concept, that of the moment of inertia (see section A.9), which in turn suggested some very nice solutions of mathematical problems [BB]. Here I solve several other mathematical problems using the center of mass.

Recall that the center of mass of a body is the body's point of balance; the body suspended on that point is in equilibrium in any orientation. Full details can be found in the appendix (section A.8).

As an interesting aside, we take it for granted, from childhood, that such a unique point of balance exists. That is, we assume that the point of balance does not depend on the orientation of the body. This assumption is true, but only if the gravitational field is constant. In variable gravitational fields, the point of balance becomes dependent on the body's orientation; here is an example. In the uniform field, a rod would balance on its geometrical center. Although the gravitational field varies even within the confines of a room, this tiny variation is drowned out by vastly stronger forces of buoyancy in the air, friction at the pivot, and so on. If, however, these annoying imperfections were somehow eliminated, we would

[1]It is striking to read Archimedes' work, available via Google Books at http://books.google.com/books?id=suYGAAAAYAAJ. Archimedes' remarkable application of the concept of center of mass to integral calculus is described in the book by Polya [P].

observe a surprising effect: the rod would balance on its center only in special orientations: vertical and horizontal. To achieve the balance at another angle we must suspend the rod off-center, at a point dependent on the angle. This effect, tiny on Earth, is clearly observable in the motion of satellites. An elongated satellite prefers to point along the radial direction.

5.2 Center of Mass of a Semicircle by Conservation of Energy

Problem. *Find the center of mass of a semicircular wire with constant linear density.*[2]

The center of mass of an object with a constant density is a purely geometrical object, referred to as the *centroid.*

This is a standard calculus exercise in using integrals—but a mechanical trick lets me bypass integration. I will rely only on the fact that

$$\lim_{\theta \to 0} \frac{1 - \cos \theta}{\theta^2} = \frac{1}{2}, \tag{5.1}$$

which can be proved using l'Hopital's rule or Taylor's expansion.[3]

Solution. Let us suspend our semicircular wire on its center, as shown in figure 5.1: just imagine gluing the wire to a weightless board and allowing the board to pivot in the vertical plane on a nail driven through the center of the semicircle. Now, let us tip the wire through a small angle θ, as shown in figure 5.1, and compute the work done in two different ways.

On the one hand, we have raised the center of mass by height $H = x - x \cos \theta = x(1 - \cos \theta)$; we thus did work $W = mgH$,

---

[2]Linear density of a wire is defined as the mass per unit of the wire's length.

[3]It can also be explained by a physical (kinematic) argument by considering the acceleration of a point moving on a circle, treating θ as time. I leave this explanation to the reader as a challenge.

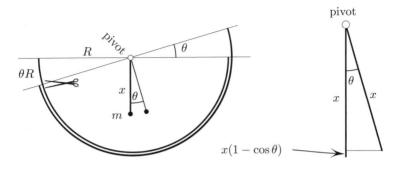

Figure 5.1. Moving the arc θR from left to right takes the same work as raising the center of mass by $x(1 - \cos\theta)$, yielding (5.3).

that is,

$$W = mg\, x(1 - \cos\theta). \tag{5.2}$$

But, on the other hand, the effect of turning was same as elevating the arc θR (figure 5.1) by the height $h = \theta R + \varepsilon$, where ε is small compared to θ in the sense that $\varepsilon/\theta \to 0$ as $\theta \to 0$. It takes work $W = \mu gh$, where μ is the mass of the arc θR. Now $\mu = \frac{\theta}{\pi} m$, since the mass of an arc is proportional to its angular size. Substituting the expressions for h and μ into $W = \mu gh$ we obtain

$$W = mg\, \frac{R}{\pi}\theta^2 + \varepsilon,$$

where ε is an error small compared to θ, in the sense mentioned before. Equating this expression for W with (5.2) yields, after canceling mg:

$$x(1 - \cos\theta) = \frac{R}{\pi}\theta^2 + \varepsilon. \tag{5.3}$$

Then, by dividing the last relation by θ^2 and setting $\theta \to 0$, we get $x/2 = \frac{R}{\pi}$ (using (5.1)), or

$$x = \frac{2R}{\pi}. \qquad\qquad \diamondsuit$$

The same method works for circular arcs of any angle, as well as for "pizza slices" (disk sectors.) The case of a "half-pizza" is

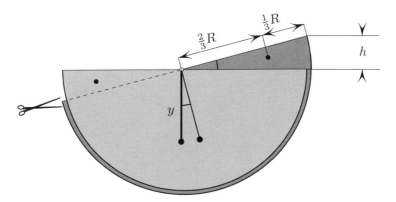

Figure 5.2. The rise of the center of mass of the sector is $2/3h$, where h is the rise of the arc.

treated next. The case of a sector of an arbitrary angle can be treated similarly (see problems at the end of this chapter).

5.3 Center of Mass of a Half-Disk (Half-Pizza)

The previous solution gives us, with almost no extra work, the center of mass of the solid semidisk. Indeed, let us repeat the argument of the preceding problem verbatim, with only one difference: instead of cutting off an arc, we cut off a thin sector from the left and place it on the right, as shown in figure 5.2.

Since our sliver of a sector approximates a triangle, we can take its centroid to be $\frac{2}{3}R$ away from the center of the circle.[4] Consequently, this centroid will rise by $\frac{2}{3}h$, where $h = \theta R + \varepsilon$ is the rise of the arc, as in the preceding problem. Thus all we have to do is to replace h in the preceding problem with $\frac{2}{3}h$; the main equation (5.3) acquires the extra factor $\frac{2}{3}$:

$$y(1 - \cos\theta) = \frac{2}{3}\frac{R}{\pi}\theta^2 + \varepsilon,$$

---

[4]We are using the fact that the centroid is at the intersection of the medians and that the medians are divided by their common point in the ratio of $1 : 2$. Thus the centroid lies on the median two-thirds of the way from the vertex of the triangle to its side.

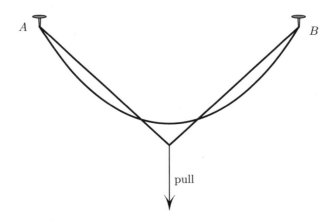

Figure 5.3. Which center of mass is higher?

where y is the unknown distance to the center (figure 5.2). Dividing by θ^2, letting $\theta \to 0$, and using (5.1) we obtain

$$y = \frac{4R}{3\pi}.$$

The centroids of the semicircle and of the semidisk are related via $y = \frac{2}{3}x$.

5.4 Center of Mass of a Hanging Chain

Problem. *A chain hangs on two nails A and B, on the same level. I grab the chain by the lowest point and pull it down. Does the center of mass of the chain move up or down?*

Solution. Of all possible shapes with A and B fixed, the hanging chain assumes the shape of least potential energy.[5] *Any* change of shape of the chain will raise its center of mass.

---

[5] Indeed, if I change the shape of the chain in any way, I have to apply force, which the chain will resist. That is, I will do positive work on the chain, thereby increasing the chain's potential energy and therefore elevating the chain's center of mass. Here I used the fact that the potential energy of an object is mgh, where m is the mass and h is the elevation of the center of mass.

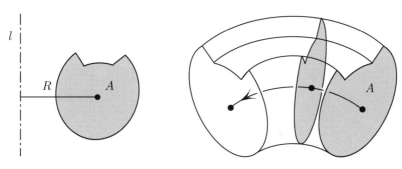

Figure 5.4. Volume of the solid.

5.5 Pappus's Centroid Theorems

Draw a line l and a region D in the plane, disjoint from each other (figure 5.4), and consider a "donut" formed by spinning D around l. What is the donut's volume? What is its surface area? Both questions are answered below.

Pappus's volume theorem. *In the setting just described, the volume of the donut of revolution of a region D around a line l is given by*

$$V = 2\pi RA,$$

where A is the area of D and where R (figure 5.4) is the distance from l to the centroid of D.

Pappus's area theorem. *The area of the donut's surface is given by*

$$S = 2\pi rL,$$

where L is the length of the curve C bounding D, and where r is the distance from l to the centroid of C. Note that in this theorem we use the centroid of the one-dimensional object: the curve, rather than the region.

The standard calculus proof of Pappus's theorem, found in most calculus books, uses integrals and their properties, and requires some

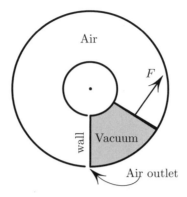

Figure 5.5. The cross section of the donut-shaped tube sealed at one end is D. A piston in the shape of D is dragged around, creating the vacuum behind it.

simple manipulation with integrals. The proof given here shows the intuitive essence of the result in its barest form.

Proof.

The mechanical system. Consider a tube bent in the shape of a donut, as in figure 5.5. The cross section of the donut tube is not circular but is exactly of the shape of D. Consider two pistons shaped as D inside the tube, as shown in figure 5.5. One piston is welded, becoming a wall, while the other can slide around the tube without friction. We also leave an air outlet as shown in figure 5.5. This bent cylinder–piston system is our mechanical analog computer that will solve the problem.

PROOF OF PAPPUS'S VOLUME THEOREM. Starting with the piston touching the wall, I grab the piston by its centroid and drag it all the way around the donut, creating a vacuum behind the piston. By computing the work done in two different ways we will obtain the statement of Pappus's theorem.

We make the following observations:

1. Holding the piston by the centroid guarantees that it will not want to pivot, which means that we need apply *no torque*, only the force F, to move the cylinder. Therefore only the force F does work.

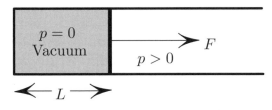

Figure 5.6. The potential energy of a bubble of vacuum equals pA.

2. The force F required to pull the piston against the atmospheric pressure p is $F = pA$, where A is the piston's area. Indeed, the pressure p is the force per unit area, and the area of the piston's face is A such units.

3. To create a volume V of vacuum against the pressure p, work $W = pV$ is required. Indeed, for a straight cylinder, the work is $W = F \cdot L = pAL = pV$. The result follows by breaking a general shape into many thin parallel cylinders.

With these remarks our proof is nearly finished. On the one hand, the work to pull the piston all the way around is force times the distance traveled by the point of application:

$$W = F \cdot 2\pi R = pA \cdot 2\pi R,$$

where R is as in the statement of the theorem. But on the other hand, by remark (3) above, work is given by the volume times the pressure:

$$W = pV.$$

Equating the two expressions yields $pV = pA2\pi R$, so

$$V = 2\pi RA.$$

This proves Pappus's volume theorem. $\diamond$

THE VOLUME THEOREM $\Rightarrow$ THE AREA THEOREM. Imagine covering the surface in question by a thin layer of paint of thickness ε, as in figure 5.7.

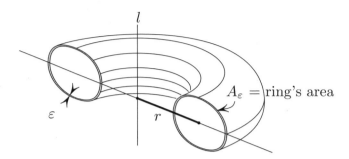

Figure 5.7. The area theorem follows from the volume theorem applied to the volume of a thin skin.

By Pappus's volume theorem, the volume of the paint is

$$V_\varepsilon = 2\pi R_\varepsilon A_\varepsilon,$$

where R_ε is the distance of the centroid of the ε-thin ring to l, and where A_ε is the ring's area. But the paint's volume V_ε is approximately the area times thickness:

$$V_\varepsilon = S\varepsilon + \cdots ; \tag{5.4}$$

the small error denoted by $\cdots$ is due to the fact that the surface is not flat. On the other hand, the centroid of the ε-ring approximates the centroid of the curve itself: $R_\varepsilon = r + \cdots$, and the ring's area $A_\varepsilon = L\varepsilon + \cdots$. Substituting all this into (5.4), dividing by ε, and letting ε approach zero results in $S = 2\pi r L$, as claimed.   ◇

5.6 Ceva's Theorem

Ceva's theorem and its converse. *Consider a triangle ABC with three points A_1, B_1, and C_1 on the sides opposite the corresponding vertices (figure 5.8). Let a, a' b, b', c, c' be the lengths as shown in figure 5.8. Ceva's theorem states that the three segments AA_1, BB_1,*

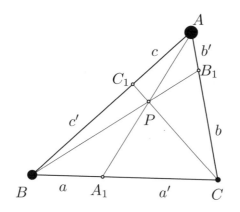

Figure 5.8. Ceva's theorem: the three Cevians AA_1, BB_1, CC_1 are concurrent if $abc = a'b'c'$.

and CC_1 *are concurrent[6] if and only if*

$$abc = a'b'c'. \qquad (5.5)$$

Proof. First, suppose that the three segments AA_1, BB_1, and CC_1 share a common point P. Let us place point masses m_A, m_B, and m_C onto the vertices of $\triangle ABC$, choosing these masses so that their center of mass lies at P. To make such a choice, we can take $m_B = a'$ and $m_C = a$, thereby placing the center of mass of (B, C) at A_1. Then we pick m_A so as to ensure that the center of mass of (A, A_1) is at P; to that end we make m_A satisfy the balance condition $m_A PA = (m_B + m_C)PA_1$. Our triangle, thus weighted, will balance on a needle point placed at P. It will then certainly balance on any line in the triangle's plane passing through P, and on the line APA_1 in particular, as figure 5.9 shows.

But since m_A lies on that line, m_C and m_B are in balance, that is,

$$m_B a = m_C a'. \qquad (5.6)$$

---

[6]That is, all intersect at one point.

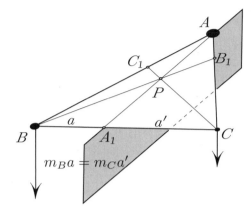

Figure 5.9. The triangle balances on the center of mass P, and hence it balances on any line through P—in particular, on AA_1.

Similarly,

$$m_C b = m_A b', \quad m_A c = m_B c'.$$

Multiplying the last three equations yields $abc = a'b'c'$.

The converse (that (5.5) implies concurrency) is easily proved by contradiction, as follows.[7] Suppose that $abc = a'b'c'$, and assume the contrary: one of the pertinent segments, say CC_1, does not pass through the intersection of the other two. A *different* segment $C\tilde{C}_1$ with $\tilde{C}_1$ on AB but $\tilde{C}_1 \neq C_1$ does pass through the intersection, and the last identity applies: $ab\tilde{c} = a'b'\tilde{c}'$. But this is in contradiction with (5.5), since $\tilde{c}'/\tilde{c} \neq c'/c$. The proof is complete.   ◇

5.7 Three Applications of Ceva's Theorem

The converse of Ceva's theorem gives an immediate proof of the following three theorems.

Theorem 1. *In any triangle the medians are concurrent.*

---

[7]See, e.g., [CG].

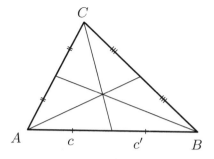

Figure 5.10. Proof of concurrency of the medians.

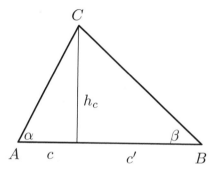

Figure 5.11. Proof of concurrency of the altitudes.

Proof. Ceva's relation (5.5) holds for the medians since $a = a'$, $b = b'$, $c = c'$, and hence the medians are concurrent by the converse of Ceva's theorem. $\diamondsuit$

Theorem 2. *The altitudes in any triangle are concurrent.*

Proof. In the notations of figure 5.11 we have $h_C = c \tan \alpha = c' \tan \beta$; similarly, $a \tan \beta = a' \tan \gamma$ and $b \tan \gamma = b' \tan \alpha$. Multiplication of the last three equations and cancellation gives $abc = a'b'c'$. By the converse to Ceva's theorem the altitudes are concurrent. $\diamondsuit$

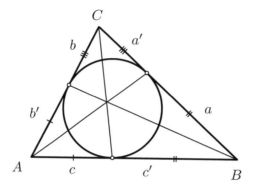

Figure 5.12. Cevians corresponding to the incircle are concurrent.

Theorem 3. *The three segments (in figure 5.12), each connecting a vertex of a triangle to the tangency point of the inscribed circle with the opposite side, are concurrent.*

Proof. Note that $b' = c$, $c' = a$, $a' = b$, since the lengths of the tangent segments from a point outside a circle to the circle are the same for both tangents. Multiplication gives $abc = a'b'c'$.   ◇

5.8 Problems

Find the center of mass of a semicircular wire by solving two subproblems, both of independent interest.

1. Find the tension of a circular rope of given mass and radius, spinning around its center with a given angular velocity (figure 5.13(*a*)), by thinking instead of a semicircular tube (figure 5.13(*b*)) with water entering at one end and exiting at the other.

2. Use the tension just found to determine the center of mass of the semicircle.

    Solution. Part 1. This is a standard problem in mechanics, but the following solution seems to be original. Let us look at all

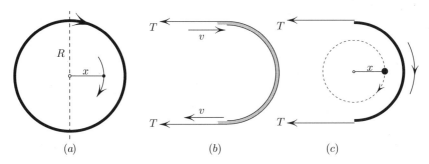

Figure 5.13. (a) A spinning flexible rope is under tension. (b) A way to find this tension by imagining water injected and ejected from a semicircular tube. (c) Using the tension to find the center of mass.

times only at the right semicircle of the rope—imagine closing the left half with a screen so that we do not have to look at it. Now the rope is "injected" on top and exits on the bottom. This is the same as water entering and exiting a tube, as shown in figure 5.13(b). The tube has to be held with some force $2T$ (where T is the tension force we are seeking). This force causes the water to reverse direction, changing speed by $v - (-v) = 2v$. During a time Δt, a column of water of length $\Delta L = v\Delta t$ will enter on top, and the same length will exit on the bottom. The net result is that the length ΔL of water changes speed by $2v$. The mass of this water is $m = \rho\Delta L = \rho v\Delta t$, where ρ is the linear density, that is, mass per unit length. By Newton's second law $F\Delta t = m\Delta v$ we have

$$2T \cdot \Delta t = m \cdot 2v.$$

After solving for T and substituting the expression for m, Δt cancels, and we obtain

$$T = \rho v^2 = \rho \omega^2 R^2,$$

where T is the tension and ω is the angular velocity.

Part 2. Let us focus attention on a material semicirclular arc of the rope. The arc stays in orbit due to two tension forces T (see figure 5.13(a)). The centripetal force $2T$ causes the centripetal acceleration of the center of mass: $M\omega^2 x = 2T$, where M is the mass of the semicircle and thus $M = \rho\pi R$. Substituting into the last equation we get $\rho\pi R\omega^2 x = 2\rho\omega^2 R^2$, so that $x = 2R/\pi$.

3. Prove that four segments, each connecting a vertex of a tetrahedron to the centroid of the opposite face, are concurrent. The tetrahedron is not assumed be be regular.

6

GEOMETRY AND MOTION

Most of the problems in this section rely on the idea of motion. The idea of motion was already used in the section on Pythagorean theorem. In section 2.4 we pointed out that the fundamental theorem of calculus can be thought of in kinematic terms. In this section I collected a few other problems, of which I like the bike problem the best. A beautiful application of the idea of motion, which allows to find the area under the tractrix with no formulas, due to R. Foote [Fo], is stated in section 6.6 as a problem. Another problem at the end of the chapter describes a way of measuring areas in the plane using a shopping cart. More on the kinematic approach can be found in the book [LS].

6.1 Area between the Tracks of a Bike

Problem. *Imagine riding a bike so that both wheels execute closed paths (figure 6.1). The front wheel never rides over the rear track, so that the distorted ring is not pinched. Show that the area of this ring does not depend on the bike's path(!).*

The area is the same whether you ride around your dining table[1] or around a block.

We assume an idealized bike: the distance b between the points of contact of the wheels with the ground is constant.[2] We will refer to b

---

[1] Bad advice involving a table was also given in section 1.3, where I suggested drilling some holes.

[2] Strictly speaking, for this to be true the front fork has to be vertical.

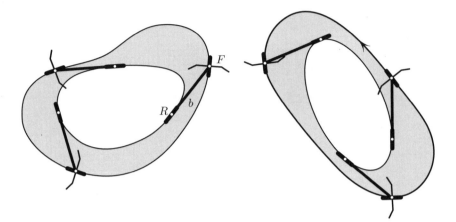

Figure 6.1. The area between the two tracks is always the same: πb^2, where b is the bike's base.

as the *base*. The problem's claim implies that the ring has area πb^2. This is the same area we would get by keeping the rear wheel fixed and pivoting the front wheel in a circle.

Solution. The bike's frame is tangent to the rear track at all times, as figure 6.1 illustrates. *The area in question is therefore swept by a tangent segment to the rear curve as this segment slides all the way around the curve. This segment has constant length b throughout its trip.*

Why is the area swept always the same? Here is the heuristic explanation.[3] Note that the rate of sweeping of area does not depend on the velocity of the segment in its own direction. Whether or not the segment has any "sliding" motion is irrelevant. But if we subtract the sliding velocity, the segment will simply rotate around its rear point R, sweeping out a circle! Hence the area of the ring is the same as that of the circle of radius $b = |RF|$ equal to the base of the bike.

[3]This is the same explanation as in the "sweeping" proof of the Pythagorean theorem in section 2.6.

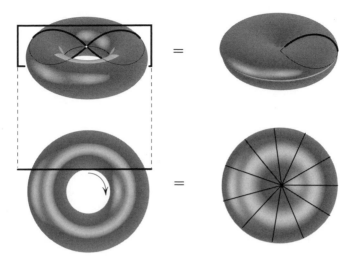

Figure 6.2. The petal sliding along the circle sweeps the same volume as the petal pivoting on its point.

6.2 An Equal-Volumes Theorem

This is a Pappus-like theorem which I stumbled upon when thinking of the bicycle problem. The intuition behind this problem is a straightforward generalization of the intuition behind the bike problem. Once we know the fact, proving it using calculus is not hard. However, I would not have discovered this fact with formulas.

Theorem. *A donut-shaped solid is obtained in the usual way by rotating a closed convex curve around an axis lying the plane of the curve (the axis does not intersect the curve). Now cut the donut with a plane parallel to the axis and tangent to the innermost parallel of the donut, resulting in a "figure eight" section, as in figure 6.2.*

Let us spin this figure eight around the line passing through the pinching point and parallel to the axis of the donut. We obtain a new solid: a differently shaped donut with a pinched hole (figure 6.2). The volume of this solid equals the volume of the original solid.

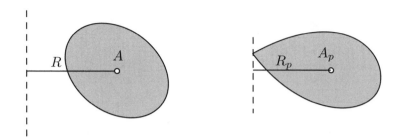

Figure 6.3. The relationship between the centroids and the areas of the two sections of a donut.

Proof. Consider one "petal" of the figure eight. As the petal slides along as shown by arrow on the left side of figure 6.2, it sweeps our given donut. In doing so, the petal executes two motions simultaneously: (i) sliding in its own plane and (ii) rotating around the line passing through the petal's pinching point as described in the theorem. If we subtract the sliding motion, we do not affect the volume swept. The resulting new solid has the same volume as the first.   ◇

Corollary. Let A be area of the disk formed by intersecting the donut with the plane through its axis, and let R be the distance from the centroid of this disk to the symmetry axis, as in Pappus's theorem. Similarly, let A_p be the area of a petal (half of the figure eight), and let R_p be the distance from the centroid of the petal to the axis as in a figure eight, (figure 6.3). Then

$$A \cdot R = A_p \cdot R_p.$$

6.3 How Much Gold Is in a Wedding Ring?

The following fact may be hard to believe at first. Having been told of it before, I could not explain *why* is it true—a formal calculation did not seem satisfying—until the "bike tracks" problem (problem 6.1) suggested the answer.

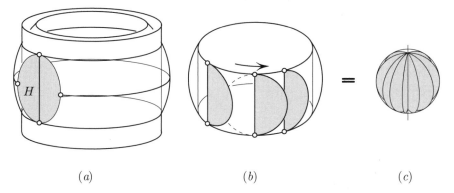

(a) (b) (c)

Figure 6.4. The volume of the ring depends *only* on its height H!

Theorem. *Figure 6.4 shows a ring whose inner surface is cylin-drical and whose outer rounded surface is spherical; the center of the sphere lies on the axis of the cylinder. The volume of such ring depends only on its height H, and is equal to $\frac{\pi}{6} H^3$.*

In particular, the ring around the Earth-sized sphere has the same volume as a wedding ring, provided the two rings have the same height (assuming both the globe and the outer surface of the wedding ring are perfect spheres).

Proof. One can prove this theorem by an explicit calculation. The following "kinematic" argument avoids formulas and makes the fact obvious. The figure tells most of the story.

Here are the details. Figure 6.4(a) shows a plane tangent to the inner cylinder. The plane cuts the sphere in a circular disk. The disk's diameter lies on a generator of the cylinder. Consider a half-disk as shown in figure 6.4(b). Now let us slide this semidisk along the cylindrical surface, as shown by the arrow, keeping the semidisk tangent to the cylinder. The moving semidisk sweeps the entire volume of the ring. But the motion of the semidisk consists of pure "sliding" in its own plane—this contributes nothing to sweeping of the volume—and of rotation around the diameter. Therefore, if we subtract the "sliding" motion, we will not change the volume swept.

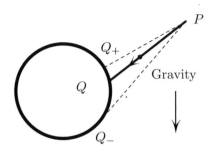

Figure 6.5. The optimal Q must lie between Q_+ and Q_-.

But the modified motion of the semidisk, shown in figure 6.4(c), sweeps a ball of diameter H. The volume of the ring is therefore the same as that of a ball of diameter H:

$$V = \frac{\pi}{6} H^3. \qquad\qquad \diamond$$

6.4 The Fastest Descent

Problem. *A circle C and a point P in the vertical plane are given. Let Q be a point on C, and consider a bead sliding along the segment PQ under the influence of gravity. The bead starts at P with zero speed. For which point Q will the travel time be minimal?*

Some remarks. Could Q_+, the point closest to P, be the answer? No. Indeed, let us move the point Q clockwise and keep track of the sliding time t_{PQ} as a function of Q. At the moment when Q passes Q_+, PQ changes with zero speed (because it just stopped shortening and is about to begin lengthening). On the other hand, the slope of PQ steepens with a positive speed. In short, at the moment when $Q = Q_+$, the length changes with zero speed, while the acceleration down PQ increases. This means that the sliding time shortens. Thus it is better to put Q below Q_+. Similarly, at the moment when Q passes $Q-$ (the point of tangency in figure 6.5) in its clockwise

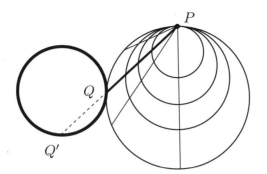

Figure 6.6. At any time $t > 0$ the beads form a circle, if they started at P with zero initial speed.

travel, the distance PQ lengthens with infinite speed, while the slope changes with finite speed. It then clearly pays to put Q above Q_-. This localizes the best point Q somewhat, but where exactly is it?

Solution. Consider a "fan" of lines through P, with a bead on each line placed at P. At $t = 0$ we release all the beads with zero initial speed, and they begin to race. At time $t > 0$ the beads form some curve shown in figure 6.6. We denote this curve by $F = F_t$ (the letter F stands for the "front," like the front of a propagating wave). As t increases, the expanding front will touch the circle at some point Q. This point of first contact gives the shortest descent time. Indeed, Q is the first point on C to be reached by any bead.

Remarkably, the front F_t turns out to be a circle, for each time t, as shown in figure 6.6! This circle passes through P with its tangent at P horizontal; the diameter of the circle is $gt^2/2$. We can now pick a circle F_t tangent to C; the "best" point Q is the point of tangency between the two circles. This answer is a bit implicit, but it is not hard to show that Q lies on the line connecting P with the lowest point Q' of the circle C.

Proving that the beads form a circle. First, recall a basic geometrical fact. Let $PP' = D$ be the diameter of a circle

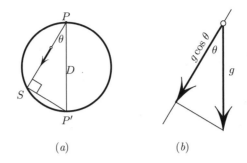

Figure 6.7. If $PS = D \cos \theta$, then S lies on the circle with the diameter $D = PP'$.

(figure 6.7), and let PS be any chord, with $\theta = \angle SPP'$. Then

$$PS = D \cos \theta. \qquad (6.1)$$

This follows from the fact that $\angle PSP' = 90°$. The converse also holds: the locus of points S satisfying (6.1) is a circle.

Now let us return to the sliding beads. Fix some $t > 0$ and consider a typical bead S at this instant. The acceleration of this bead is

$$a = g \cos \theta,$$

as figure 6.7(b) illustrates. The distance this bead will travel in time t is therefore

$$PS = at^2/2 = (gt^2/2) \cos \theta = D \cos \theta,$$

where $D = gt^2/2$ is the distance of free fall. We showed that at time t, every bead S satisfies (6.1). According to the preceding geometrical remark, all beads S lie on the circle of diameter $D = gt^2/2$ with the top point at P, as claimed.

6.5 Finding $\frac{d}{dt} \sin t$ and $\frac{d}{dt} \cos t$ by Rotation

Consider a point P moving on a unit circle with unit speed in the counterclockwise direction, starting at the point P_0 on the x axis at $t = 0$. The arc $P_0 P$ therefore has length t, that is, $\angle P_0 O P = t$. By

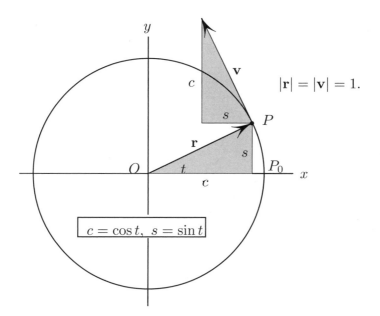

Figure 6.8. $s = \sin t$, $c = \cos t$.

the definition of sine and cosine, the position vector

$$\overline{OP} = \langle \cos t, \sin t \rangle.$$

The velocity is given by the derivative of the position:

$$\mathbf{v} = \left\langle \frac{d}{dt} \cos t, \frac{d}{dt} \sin t \right\rangle.$$

On the other hand, since $|\mathbf{v}| = 1$ (by assumption), the shaded right triangles in figure 6.8 have congruent hypotenuses (of length 1). Moreover, since $\mathbf{v} \perp \overline{OP}$, all the corresponding sides of the two triangles are perpendicular and hence the corresponding angles are congruent. The shaded triangles are therefore congruent, and thus

$$\mathbf{v} = \langle -\sin t, \cos t \rangle.$$

Comparing with the previous equation we conclude $\frac{d}{dt} \cos t = -\sin t$, $\frac{d}{dt} \sin t = \cos t$.

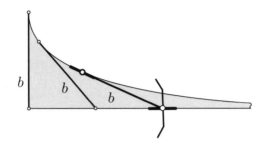

Figure 6.9. What is the area under the tractrix?

6.6 Problems

1. (Foote [Fo]) The tractrix is, by definition, the curve formed by the rear track of a bike when the front track moves in a straight line. More precisely, a curve is called a tractrix if there exists a length b and a straight line such that the segment of any tangent lying between the curve and the line has length b. In figure 6.9 the semi-infinite tractrix is sketched. What is the area of the shaded infinite figure?

   Answer: $\frac{1}{4}\pi b^2$.

2. While waiting in a long checkout line in a supermarket, I am tracing the outline of a floor tile with a front wheel of my shopping cart. After one round trip of the front wheel, the rear wheels end up in a new place. That is, the cart has pivoted around the front wheel through some angle θ. What is the approximate value of θ, given any desired information (short of the answer)? The area of the tile is A, the distance between the front and the rear wheel is b. Assume that b is much longer than the side of the tile. To further simplify things, assume that the rear wheel aims exactly at the front wheel (in an actual cart this is not so; the wheels form a trapezoid).

   Answer: $\theta \approx A/b^2$. For much more on this problem, see [Fo], [LW], and references therein.

7

COMPUTING INTEGRALS USING MECHANICS

The first two problems in this section are easy to do with calculus and without mechanics. The point here is to illustrate how the "thinking" that the calculus machinery does for us can sometimes be done by a mechanical "analog computer."

7.1 Computing $\int_0^1 \frac{x\,dx}{\sqrt{1-x^2}}$ by Lifting a Weight

A weight $P = 1$, mounted on a frictionless vertical track, hangs on a string of length 1. The string is vertical initially. As the top end of the string is moved horizontally from its initial position, the weight slides upward along the vertical line. In changing the displacement x of the top end of the string from $x = 0$ to $x = 1$, we do work $W = \int_0^1 F(x)dx$, where $F(x)$ is the force required to hold the end of the string at x, as in figure 7.1.[1]

On the other hand the same work goes into raising the weight P by height 1, so that $W = P \cdot 1 = 1$, and we conclude $\int_0^1 F(x)dx = 1$. Now I claim that $F(x) = \frac{x}{\sqrt{1-x^2}}$. Let O be the point of intersection of two perpendiculars, as shown in figure 7.2.

The sum of torques upon the string AB relative to O vanishes, which gives

$$OB \cdot F = OA \cdot P,$$

---

[1] The value $x = 1$ is not reachable, since it requires infinite force; we are dealing with an improper integral.

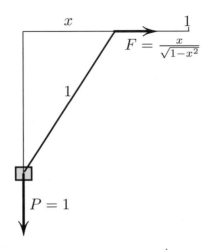

Figure 7.1. The work done by raising the weight is $\int_0^1 F(x)dx = P \cdot 1$.

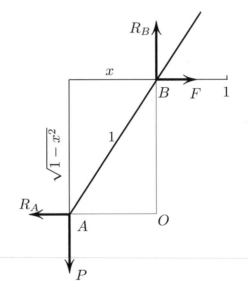

Figure 7.2. The normal reaction forces $-F$ and $-P$ have zero torque relative to O.

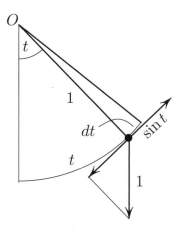

Figure 7.3. The work required to pull the weight from the bottom to angle x is $\int_0^x \sin t\, dt$.

and since $OA/OB = x/\sqrt{1-x^2}$ and $P = 1$, we obtain $F(x) = \frac{x}{\sqrt{1-x^2}}$, as claimed. We showed that

$$\int_0^1 \frac{x}{\sqrt{1-x^2}}\,dx = 1.$$

7.2 Computing $\int_0^x \sin t\, dt$ with a Pendulum

Consider a pendulum: a point mass of weight $P = 1$ on a stick of length 1, pivoting on a hinge O. Force $\sin t$ is needed to hold the pendulum at an angle t to the vertical, as seen in figure 7.3. Since the radius of the rod is 1, the angle t measures the length along the circle. Therefore, the work required to move the bob from t to $d + dt$ is $\sin t\, dt$, and the total work needed to pull the bob from the bottom $t = 0$ to $t = x$ is $\int_0^x \sin t\, dt$. On the other hand, the change in potential energy is (weight) $\cdot$ (height) $= 1 - \cos x$ (figure 7.4). Equating two expressions for the same energy gives

$$\int_0^x \sin t\, dt = 1 - \cos x.$$

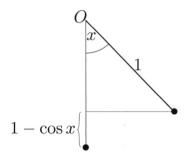

Figure 7.4. Potential energy for the deflection angle x is $1 - \cos x$. The weight is 1.

7.3 A Fluid Proof of Green's Theorem

This short discussion makes the concept of divergence and Green's formula

$$\int_C \mathbf{F} \cdot \mathbf{n} \, ds = \iint_D \operatorname{div} \mathbf{F} \, dA \qquad (7.1)$$

seem almost trivial. This heuristic discussion uses the concepts of the dot product and the line integral.

We are given (i) a planar vector field $\mathbf{F} = \mathbf{F}(x, y)$ and (ii) a curve C bounding a planar region D. The key is to think of $\mathbf{F}$ as the velocity field of an imaginary planar gas, and to let the domain D flow according to the velocity field $\mathbf{F}$, obtaining a new domain $D(t)$ at time t with $D(0) = D$. Let $A(t) = \operatorname{area}(D(t))$.

Dividing the domain D into a large number N of small pieces D_n, $1 \leq n \leq N$, we split up its area:

$$A(t) = \sum A_n(t).$$

Differentiating by t at $t = 0$, we get (using the notation $\dot{} = d/dt$):

$$\dot{A}(0) = \sum \dot{A}_n(0). \qquad (7.2)$$

But the rate of change $\dot{A}_n(0)$ of the area of each small piece should be approximately proportional to its area:

$$\dot{A}_n(0) = k A_n(0) + \varepsilon, \qquad (7.3)$$

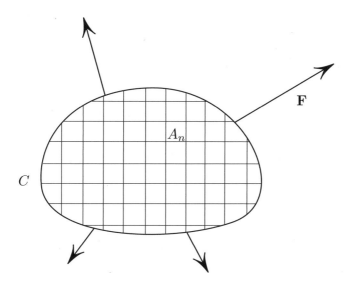

Figure 7.5. The initial domain is carried by the flow **F**.

where ε is a small error: $\varepsilon/A_n(0) \to 0$ if $A_n(0) \to 0$. *The coefficient of proportionality k, in the limit of the area shrinking to a point, is called the* **divergence** *of* **F** *at that point.* That is, one defines

$$\operatorname{div} \mathbf{F}(x, y) = \lim \frac{\dot{A}(0)}{A(0)},$$

where the limit is taken as the domain shrinks toward the point (x, y). The coefficient k tells us how fast the area expands (or contracts) per unit area at a point.[2] Thus k is indeed a measure of the divergence of the gas particles.

Now substituting (7.3) into (7.2) we obtain

$$\dot{A}(0) = \sum \operatorname{div} \mathbf{F}(x_n, y_n) A_n(0) + \text{small error},$$

---

[2]It is a simple matter to obtain the formula $\operatorname{div} \mathbf{F} = \operatorname{div} \langle P, Q, \rangle = \frac{\partial P}{\partial x} + \frac{\partial Q}{\partial y}$ from this definition.

where (x_n, y_n) is a point inside the nth "cell." In the limit of infinitely fine partition we obtain

$$\dot{A}(0) = \iint_R \operatorname{div} \mathbf{F} \, dA.$$

Now the rate of change of area of a moving region is the integral of the normal component of the speed of its boundary: $\dot{A}(0) = \int_C \mathbf{F} \cdot \mathbf{n} \, ds$ (here $\mathbf{n}$ is the outward unit normal to C). This completes the sketch of the proof of Green's theorem (7.1).

8

THE EULER-LAGRANGE EQUATION VIA STRETCHED SPRINGS

8.1 Some Background on the Euler-Lagrange Equation

This short chapter contains a purely mechanical interpretation of the Euler-Lagrange functional as the potential energy of an imaginary spring. This interpretation makes for an almost immediate derivation of the Euler-Lagrange equations and gives a very transparent mechanical explanation of the conservation of energy. Moreover, each individual term in the Euler-Lagrange equation acquires a concrete mechanical meaning.

Here is some motivation for the reader not familiar with the Euler-Lagrange equations.

A basic problem of the calculus of variations is to find a function $x(t)$ which minimizes an integral involving x and its derivative $\dot{x}$:

$$\int_0^1 L(x(t), \dot{x}(t))\, dt, \qquad (8.1)$$

where L is a given function of two arguments, and where the boundary values of x are given: $x(0) = x_0$, $x(1) = x_1$.

To give the flavor of the rest of the chapter, let us look at an example: minimize

$$\int_0^1 \frac{\dot{x}^2}{2}\, dt \qquad (8.2)$$

subject to $x(0) = 0$, $x(1) = a$. How to do it? Instead of using the standard theory, I will illustrate in this particular example how mechanics solves this problem. The same approach applies in general, as shown later.

I want to interpret the integral (8.2) as the potential energy of a spring. To do this, let us imagine a spring laid along the x axis. We interpret $x(t)$ as the position of a particle of the spring, where t is the parameter labeling the particles of the spring. We hold the ends of the spring at $x = 0$ and at $x = a$: $x(0) = 0$, $x(1) = a$. We also assume that the spring is elastic, in the sense that the tension force of the spring is given by Hooke's law: $T(x) = dx/dt = \dot{x}$. Note that I can stretch the spring by different amounts in different places, so that $T(x(t)) = \dot{x}(t)$ can vary with t. I claim that (8.2) is precisely the potential energy of such a spring (where each particle t is held at $x = x(t)$, perhaps by force). Postponing the proof to the next paragraph, we are done: if a particular function $x(\cdot)$ delivers a minimum of (8.2), then the corresponding configuration of the spring is in equilibrium, and thus the tension throughout the spring is constant:[1] $\dot{T} = 0$, or $\ddot{x} = 0$, so that x is a linear function; the boundary conditions yield $x = at$ as the solution.

It remains to explain why (8.2) is the potential energy. Consider a small element of the spring corresponding to the interval $(t, t + dt)$. Let us start with the element unstretched, and then stretch it from its zero length to the final length $x(t + dt) - x(t)$. As we stretch it, the force changes from 0 to $\dot{x}(t)$ linearly with the distance, so that the average force we apply is $\frac{1}{2}\dot{x}$. On the other hand, we cover the distance $x(t+dt) - x(t) = \dot{x}dt$, for the total work $\frac{1}{2}\dot{x}^2 dt$, as claimed!

Although Archimedes could have discovered the approach outlined above, and no doubt could have generalized it, the problem was not posed to him. Instead, Euler and Lagrange solved the problem by different methods. They found that if a function $x = x(t)$ gives a minimum to the integral (8.1), then it must satisfy the condition

$$\frac{d}{dt}L_{\dot{x}}(x, \dot{x}) - L_x(x, \dot{x}) = 0; \qquad (8.3)$$

the subscripts here denote partial derivatives. Note that our simple case follows from this result. Indeed, if $L = \dot{x}^2/2$, then $L_{\dot{x}} = \dot{x}$,

---

[1] Since no other forces act on a spring.

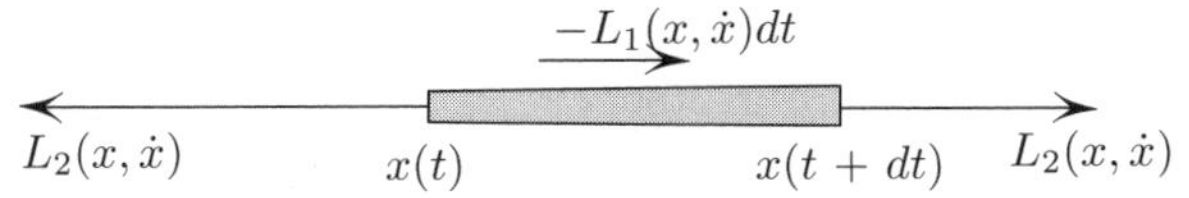

$$-L_1(x, \dot{x})dt$$

$$L_2(x, \dot{x}) \qquad x(t) \qquad\qquad x(t + dt) \quad L_2(x, \dot{x})$$

Figure 8.1. The potential force balances the tensions: this is a mechanical interpretation of the Euler-Lagrange equation (8.3).

$L_x = 0$, and (8.3) becomes $\ddot{x} = 0$, as we discovered by a naive mechanical argument. The same argument turns out to extend to the general case, leading to (8.3) and to its mechanical interpretation.

Since the time of Euler and Lagrange, the standard way to derive (8.3) is to use infinitesimal variations [GF]. Our goal here is to point out that an approach that Archimedes could have used gives an alternative derivation, and, what is more, provides a concrete mechanical interpretation to the individual terms in the equation. In a nutshell, (8.3) can be viewed as the condition that a hanging "slinky" is in equilibrium.

I should point out that the arguments here are not rigorous; the goal is rather to show that the theory has a palpable mechanical interpretation.

8.2 A Mechanical Interpretation of the Euler-Lagrange Equation

Let us imagine an idealized spring, like a heavy rubber band or a slinky, treated as an infinitely thin line, laid along an x axis. The particles of our spring are labeled by a parameter $t \in [0, 1]$, so that $x(t)$ is the coordinate of the corresponding particle.

Let t measure the mass of the slinky, so that the segment $[x(t), x(t + dt)]$ has mass dt. We now endow the Euler-Lagrange integral (8.1) with the meaning of total potential energy of the string.

1. $L(x, 0)$ is the potential on the line; in other words, a *point* mass dm located at x has potential energy $L(x, 0)dm$. By the definition of the potential energy, the corresponding force upon this mass is

$$-L_x(x, 0)dm.$$

2. The slinky satisfies an analog of Hooke's law: the tension is $T(x, \dot{x}) = L_{\dot{x}}(x, \dot{x})$. For example, in the most important case of $L = \frac{1}{2}m\dot{x}^2 - V(x)$ we have $T = m\dot{x}$, the linear Hooke's law. But in general the tension depends not only on the stretching $\dot{x}$ but also on the location x.

$\int L(x, \dot{x}) \, dt =$ **slinky's total energy.** Indeed, consider a short segment $[x(t), x(t + dt)]$ of the slinky, of mass dt. The potential energy consists of two parts: first, the potential energy of a *point* mass dt, given by $L(x(t), 0)dt$, and second, the work required to stretch this point mass dt to become our segment; this work is given by

$$\int_{x(t)}^{x(t+dt)} T\left(x, \frac{\sigma - x(t)}{dt}\right) d\sigma = \int_0^{\dot{x}} L_2(x, s)ds \, dt$$

$$= (L(x, \dot{x}) - L(x, 0)) \, dt,$$

where $\frac{\sigma - x(t)}{dt} = s$. The segment's total energy, potential + stretching, is then $L(x, \dot{x})dt$. The total energy of the slinky is then the Euler-Lagrange integral (8.3), as claimed.

8.3 A Derivation of the Euler-Lagrange Equation

If a function $x(t)$ minimizes the Euler-Lagrange integral, then the corresponding slinky is in equilibrium. The difference in tensions at the two ends of an infinitesimal segment is then in balance with the potential force

$$L_{\dot{x}}(x, \dot{x})\big|_t^{t+dt} = \int_t^{t+dt} L_x(x, \dot{x})d\tau.$$

Dividing both sides by dt and taking the limit as $dt \to 0$ leads to the Euler-Lagrange equation.

To summarize, we have endowed each term in the Euler-Lagrange equation with a mechanical interpretation: $L_{\dot{x}}$ is the tension, $\frac{d}{dt}L_{\dot{x}}$ is the resultant of tensions per unit mass, and L_x is the potential force per unit mass.

8.4 Energy Conservation by Sliding a Spring

The Euler-Lagrange equation (8.3) has a property hidden to a naked eye. It turns out that

$$L - \dot{x}L_{\dot{x}} = \text{constant} \qquad (8.4)$$

for any solution $x = x(t)$ of the Euler-Lagrange equation. What is the meaning of this quantity[2] in our "slinky" model? The answer is hidden in this obvious observation: Since the slinky is uniform, that is, its energy density $L(x, \dot{x})$ doesn't depend on t explicitly, we know that

The energy contained in a segment $a \leq x \leq b$ does not change

if we slide the slinky so that $x(t)$ moves to $x(t + c)$, (8.5)

where c is a constant. Let us now translate the obvious (8.5) into the nonobvious (8.4). Let us slide the slinky to the right with $c = dt$, by letting the mass dt slide into $[a, b]$ at $x = a$ and by pulling the same mass out at $x = b$. The change of energy inside $[a, b]$ is zero by (8.5) on the one hand. On the other hand, this change is composed of the energy of added mass minus the energy of removed mass **and** of the work done by the tension forces on the two ends. The added/removed energy is

$$Ldt \left.\right|_{t=t_a}^{t=t_b},$$

while the work done by tension forces at each end is

$$L_2 dx|_{t=t_a} - L_2 dx|_{t=t_b} = -\dot{x}L_2 dt \left.\right|_{t=t_a}^{t=t_b}.$$

Setting the sum of these two energies to zero and dividing by dt gives

$$L - \dot{x}L_2 \big|_{t=t_a}^{t=t_b} = 0,$$

proving the constancy of the energy (8.4)!

---

[2]In the special case when L is the difference of the potential and the kinetic energies, $L - \dot{x}L_{\dot{x}}$ turns out to be the total energy, i.e., the sum of the two energies.

9

LENSES, TELESCOPES, AND HAMILTONIAN MECHANICS

The central point of this chapter is a very simple hand-waving (in a literal sense) argument in mechanics in section 9.3. This simple mechanical argument has rather unexpected consequences in mathematics and in optics.[1] Thanks to the mechanical interpretation, some of these consequences, usually discussed only in graduate courses, become much more accessible.

Here is the plan of the chapter. Section 9.1 contains the background; sections 9.3 and 9.2 describe the mechanical system and give a mechanical proof of a geometrical theorem on area preservation. Section 9.7 connects the mechanical/geometrical problem with an optical one, and the last section (9.8) explains the functioning of telescopes and other optical devices via the "uncertainty principle," which in turn came from mechanics.

Here are the chapter's highlights:

1. Area-preserving mappings arise naturally in mechanics (sections 9.2 and 9.3).
2. A table of analogies between mechanics and area-preserving maps (section 9.5).
3. The uncertainty principle—a classical counterpart (section 9.6).
4. The working of telescopes, binoculars, and other optical devices is explained via the "uncertainty principle" (section 9.8).

---

[1] This is not the first time that we see something trivial with nontrivial consequences.

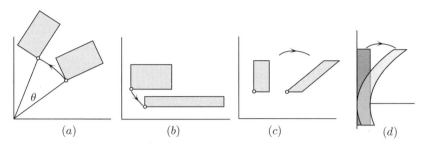

Figure 9.1. Examples of area-preserving mappings.

9.1 Area-Preserving Mappings of the Plane: Examples

A mapping of the plane is, by the definition, a function which to every point $z = (x, y)$ in the plane assigns a new point

$$\varphi(z) = (f(x, y), g(x, y)). \tag{9.1}$$

A mapping is said to be *area preserving* if the area of any set equals the area of its image under the mapping. Simplest examples of the area preserving mappings include

1. A rotation: to every point (x, y) the map assigns the point rotated through a given angle θ. This point is given by $x \cos\theta - y \sin\theta$, $x \sin\theta + y \cos\theta$), as in figure 9.1($a$).
2. A hyperbolic rotation: stretching by a constant factor λ in one direction and contraction by the same factor in another; for instance, $(x, y) \mapsto (\lambda x, \lambda^{-1} y)$, as in figure 9.1($b$).
3. A parabolic rotation $(x, y) \mapsto (x + ay, y)$, or a shear map, as in figure 9.1(c).
4. A parabolic shear: $(x, y) \mapsto (x + y^2, y)$, as in figure 9.1($d$).

9.2 Mechanics and Maps

A remarkable connection between geometry and mechanics is a recurring pattern in this book. I will describe one aspect of this connection in a maximally simple form, stripped of most terminology and of all technicalities. Using mechanics, we will obtain an

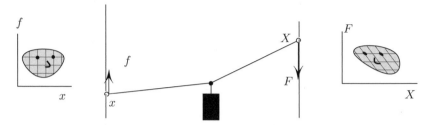

Figure 9.2. Mechanics leading to geometry: The mechanical system gives rise to a mapping ψ from the (position, force) pair (x, f) on the left to a similar pair (X, F) on the right.

Figure 9.3. The potential energy of a zero-length spring is a generating function the shear map.

explanation of why telescopes, microscopes, and binoculars magnify objects.

Here is this geometry-mechanics connection in a nutshell. Consider a simple mechanical system, such as the one in figure 9.2.[2] Each of the two rings can slide frictionlessly along its line. Imagine me holding the two rings, with everything at rest. The position x of the left ring and the force f with which it pulls my left hand determine[3] the position X of the right ring and the force F with which it tries to move. We thus have a mapping

$$\psi = (x, f) \mapsto (X, F).$$

Here comes the remarkable part: the mapping is area preserving, no matter how complicated the mechanical arrangement between the two rings might be. We can change ψ by adding springs, pulleys, more weights, but we cannot change its property of area preservation. (For other examples, see figures 9.3 and 9.7.)

The explanation is very simple and is given in the next section.

[2]More examples are given in problems in section 9.9.
[3]Under some technical assumptions which are described later.

9.3 A (Literally!) Hand-Waving "Proof" of Area Preservation

Here is the proof that the map $\psi : (x, f) \mapsto (X, F)$ preserves area. Referring to figure 9.2, imagine me holding the two rings so that the assembly is stationary. To keep everything still I have to apply the force $(-f)$ with my left hand and the force $(-F)$ with my right hand.

Now here comes the hand-waving part: very slowly (so as not to excite any oscillations) I move my hands in some arbitrary but cyclic fashion, bringing the rings back to their original position. I end up doing zero work: $\oint (-f)\, dx + \oint (-F)\, dX = 0$, or

$$\oint f\, dx + \oint F\, dX = 0, \qquad (9.2)$$

where the first term is the work done by my left hand and the second term is the work done by my right hand. Thus far this is a completely obvious mechanical statement. But let us translate it into a geometrical one. When I move my hands, the points (x, f) and (X, F) describe closed paths c and C respectively. Moreover, the second curve is the image of the first: $C = \psi(c)$. But note that the two terms in (9.2) are precisely the areas of c and C! The work done by each of my two hands has a geometrical· meaning of the area! The sum of two areas is zero; in other words, up to the change in sign, the mapping ψ preserves the area. This change of the sign means the change in orientation: the image of the left glove will look like a right glove.

A cosmetic touch-up. To avoid dealing with negative areas, and for historical reasons, let us introduce $Y = -F$; for the sake of uniformity in notation let us also rename $f = y$ (no sign change here). Then (9.2) turns into

$$\oint y\, dx = \oint Y\, dX, \qquad (9.3)$$

which means that the mapping $\varphi = (x, y) \mapsto (X, Y)$ preserves area, this time including the sign.

This property (and its higher dimensional analogs) comes up in different guises and it has profound consequences in dynamics and optics. In fact, this observation led to a very active area of current research in symplectic topology [HZ].

9.4 The Generating Function

The "proof" of area preservation was not mathematically rigorous, since I didn't even define the mapping φ precisely, saying only that it is defined by a vaguely described mechanical system. Here is a way to make all this precise. All we really need from figure 9.2 is the potential energy $P = P(x, X)$ of the system. By the definition of the potential energy,[4] we have the forces

$$\begin{cases} f = -\dfrac{\partial}{\partial x} P(x, X) \\ F = -\dfrac{\partial}{\partial X} P(x, X). \end{cases} \tag{9.4}$$

This is the precise definition of the map ψ. The map is defined by specifying a function P, called the *generating function*. The cosmetically changed map $\varphi = (x, y) \mapsto (X, Y)$ is given by changing the sign of F, as mentioned before, and by renaming f into y:

$$\begin{cases} y = -\dfrac{\partial}{\partial x} P(x, X) \\ Y = \dfrac{\partial}{\partial X} P(x, X). \end{cases} \tag{9.5}$$

Example. Consider a simple quadratic function $P(x, X) = \frac{1}{2}k(X - x)^2$. It can be interpreted as the potential energy of the zero-length spring whose ends are held at the points x, X (figure 9.3), with Hooke's constant k. Equations (9.5) give $X = x + ky$, $Y = y$; this is exactly the shear map defined in section 9.2, with $a = k$!

---

[4]Recall that if $P(x)$ is the potential energy, then the force is $-P'(x)$.

Interestingly, the strength of the shear, a geometrical property, is interpreted as Hooke's constant, a mechanical attribute. Note also that $X - x = ky$ is just Hooke's law for the zero-length spring, while $Y = y$ means that we pull the two ends of the spring with equal forces!

9.5 A Table of Analogies between Mechanics and Analysis

Mechanics	**Analysis**
The potential energy $P(x, X)$	The generating function $P(x, X)$
The forces $f = -\frac{\partial}{\partial x} P(x, X)$, $F = -\frac{\partial}{\partial X} P(x, X)$	The momenta $y = -\frac{\partial}{\partial x} P(x, X)$, $Y = \frac{\partial}{\partial X} P(x, X)$
The work done by the left(right) hand: $\oint f \, dx \ (\oint F \, dX)$	The area of the preimage (the image): $\oint y \, dx \ (\oint Y \, dX)$
Zero net work done: $\oint f \, dx + \oint F \, dX = 0$	Area is preserved: $\oint y \, dx = \oint Y \, dX$

Higher dimension. The area-preservation property (9.3) can be generalized to mappings in higher dimensions by allowing both x and y to be in the n-dimensional space ($n \geq 1$): $x \in \mathbb{R}^n$, $y \in \mathbb{R}^n$, so that $(x, y) \in \mathbb{R}^{2n}$. Then $y \, dx$ in (9.3) must be understood as the dot product: $y \, dx = \Sigma y_k \, dx_k$. The mapping $\varphi(x, y) \to (X, Y)$ of $\mathbb{R}^{2n}$ satisfying the property (9.3) is referred to as **symplectic**.

The mechanical interpretation of figure 9.2 becomes only simpler in the case of higher dimension $n = 3$: I no longer have to restrict the rings to lines, so that now $x, f, X, F \in \mathbb{R}^3$. Now (x, f) lies in $\mathbb{R}^6$ and we have a mapping $\varphi : \mathbb{R}^6 \to \mathbb{R}^6$. This mapping assigns to the (position, force) pair of my left hand the (position, force) pair of my right hand. The mapping is automatically symplectic (after a sign change as described in the last theorem). When two children are holding a skipping rope by its ends, they are dealing with a symplectic map of $\mathbb{R}^6$ (assuming the rope is still).

How general is the mechanical interpretation? One can show that, in fact, any symplectic map, reasonably nondegenerate, is a composition of maps realizable by mechanical systems similar to the ones just described.

9.6 "The Uncertainty Principle"

The quantum mechanical uncertainty principle states, roughly speaking, that the more certain we are about the position of a particle, the less certain we can be about its velocity.

The area-preservation property can be viewed as a classical mechanical analog of the uncertainty principle. Figure 9.4 shows an area-preserving map which squeezes the width of a region in the x direction from 1 to $\varepsilon \ll 1$. We can think of this squeezing as the gain of information about x, since the range of possible x values gets narrowed. But to be area preserving, the map must stretch in the y direction. This stretching means that the range of Y values is large, so that we lost information about Y. In summary, having gained information about X we lost information about Y; these variables can be viewed as analogs of the quantum mechanical position and momentum, and in fact they often arise in classical mechanics as position and momentum.

Problem. *Assume that mapping in figure 9.4 squeezes the region as illustrated. Show that the range of Y values in the image of the region of area A is at least A/ε.*

Telescopes magnify objects thanks to this uncertainty principle, as I explain later (section 9.8).

9.7 Area Preservation in Optics

Consider a ray passing through an optical device—a scope, a telescope, a binocular, or a microscope—as in figure 9.5.

Let x and X denote the coordinates of the intersection of the ray with two parallel axes, x and X. The time of travel between

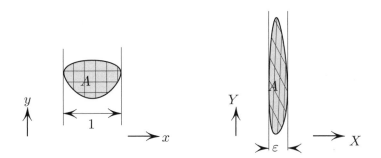

Figure 9.4. Squeezing the x range expands the y range, according to area preservation.

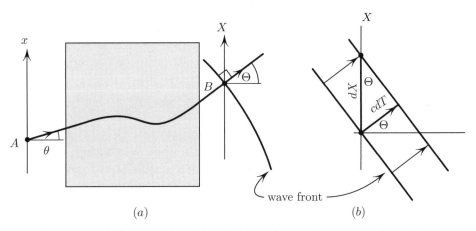

Figure 9.5. (*a*) The travel time $T(x, X)$ defines the mapping from (x, θ) to (X, Θ) via (9.6). (*b*) Proof of (9.6).

these lines is denoted by $T(x, X)$. Let $L(x, X) = cT(x, X)$, where $c = $ constant is the speed of light in the air.[5] Let $y = \sin \theta$, where θ is the angle between the ray and the horizontal axis perpendicular to the x axis in figure 9.5(*a*). Similarly, we define $Y = \sin \Theta$. We will show that the entry data (x, y) for the ray are related to the

[5]Thus L has the dimension of the distance; note that L is greater than the actual length of the ray since part of the time is spent in the glass where the light is slower, so that T is greater than the time the same path would take in the air. Actually we can choose the units so that $c = 1$, in which case we would have $T = L$.

exit data (X, Y) via

$$\frac{\partial}{\partial x}L(x, X) = -y,$$

$$\frac{\partial}{\partial X}L(x, X) = Y.$$
(9.6)

This is exactly the same relationship as (9.5) in our example from mechanics, figure 9.2! We proved by a mechanical "hand-waving" argument (section 9.3) that the mapping $(x, y) \mapsto (X, Y)$ is area preserving. This is a fundamental property of any optical device. We (literally) see the effects of this area preservation when the light travels from, say, a TV screen, through the eyeglasses, through cornea, and onto the retina. I will describe one fascinating manifestation of this area preservation in the next section.

*A **mechanical analogy.*** The table of analogies (section 9.5) extends from mechanics to optics: the potential $P(x, X)$ corresponds to the length $L(x, X) = cT(x, X)$; the force $F = \frac{\partial}{\partial X}P(x, X)$ corresponds to $\sin \Theta$.

Proof of (9.6). Figure 9.5 shows a wavefront of rays starting at A. By definition, all the points on a wavefront are equidistant from A in the sense that all take the same *time* to get to from A—"isochronous" would be a better word.[6]

1. It turns out that the *wavefront is always perpendicular to the ray*, provided the medium is isotropic,[7] as we assume all the lenses in our optical device to be. One can derive this orthogonality from Huygens's principle; because of the isotropy the infinitesimal sets reachable in time dt from a point (these sets are called the indicatrices) are spherical rather than elliptical. Further details can be found in [ARN]. In Euclidean geometry, the radii are perpendicular to their circles; in the same way, and for essentially the same reason, our rays are perpendicular to their

---

[6]Since the time is referred to as the *optical length*, we can say that the front originating at A is an optically equidistant set, i.e., a circle in the sense of optical length.

[7]This means that the the speed of light at each point does not depend on the direction.

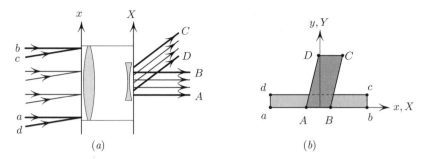

Figure 9.6. If the beam narrows, then the angle between two beams widens. This is perceived as magnification. If the beam narrows, say, four times, then the device magnifies by the factor of 4.

wavefronts—recall that the wavefronts are defined as circles with the time playing the role of distance.[8]

2. Consider two consecutive positions of the wavefront, a short time dT apart (figure 9.5). The front displaces by cdT, and the intersection with the X axis moves by $dX = cdT/\sin\Theta$. Thus $cdT/dX = dL/dX = \sin\Theta = Y$.   ◇

Remark. Formula (9.6) depends on the fact that the angles θ, Θ are measured in the air.

9.8 Telescopes and Area Preservation

Consider an optical instrument, such as a telescope or a binocular (the latter is shown schematically in figure 9.6). For us the instrument is a black box: we don't know or care what lenses or mirrors are inside, or how many of them. All we are given is this:

A parallel beam of rays is converted into a narrower *parallel beam.*

Just this fact alone implies that the optical device magnifies objects! Why? Here is a sketch of the answer, with details to follow. In the preceding section we considered the mapping

[8] All this is a subject of differential geometry; a good discussion can be found in, e.g., [DO].

$(x, y = \sin\theta) \mapsto (X, Y = \sin\Theta)$, which assigns the entry data of the ray to the exit data. We showed that the mapping is area preserving. Now, the narrowing of the beam implies that the mapping squeezes a thin rectangle $abcd$ in the x direction. The area preservation then forces a compensatory stretching in the y direction. But the stretching in the y direction manifests itself as the image magnification.

We will show, in fact, that the magnification factor equals the ratio of the beams' widths, assuming that the angles with the optical axis (i.e., the axial line of symmetry) are small.

Here is a more detailed explanation of the above. In the discussion that follows we will deal with small angles, which will allow us to approximate $y = \sin\theta$ by θ and $Y = \sin\Theta$ by Θ.

1. Each ray—say, cC in figure 9.6—is specified by the point (x, y) in the plane, where x is the coordinate of the entry and y is approximately the angle of the ray with the optical axis.

2. The magnification factor of the binocular is the ratio θ_2/θ_1, where θ_1 is the angle between two parallel beams entering the binocular, and θ_2 is the angle between the exiting beams (figure 9.6). First, note that a parallel beam is perceived by our eye as a point, since all parallel rays focus on one "pixel" of our retina (assuming we focus our eyes on infinity and have perfect vision). Now, imagine we are looking at a distant ship, with one parallel beam ab (figure 9.6) coming from a point[9] on the stern and with another beam dc coming from a point on the bow. To a naked eye the stern and the bow appear very close because the beams come in at a small angle; I barely need to rotate my eye to look from one to another. But the converted beams AB and CD form a much larger angle, so that the stern and the bow will appear much farther apart—in fact, precisely θ_2/θ_1 times farther.

3. The incoming ab beam in figure 9.6(a) corresponds to the segment ab in the same figure (b); the slanted incoming beam in (a) corresponds to the segment cd in (b). The beams whose angles

[9]The rays radiating from a point are not strictly speaking parallel, but the ship is far away and we can treat them as parallel.

and positions are between these two extremes form the interior of the rectangle *abcd*. The horizontal segment *ab* maps to a shorter segment AB by the beam-narrowing property. The length of a horizontal segment is the width of the corresponding parallel beam. Hence the ratio of the lengths $ab/AB = \lambda$, the same as the ratio of the beams' widths. By linearity, every horizontal segment in *abcd* (i) shortens by the same factor λ and (ii) remains horizontal. The height of *abcd* is $\sin\theta_1$, while the height of $ABCD$ is $\sin\theta_2$. The areas of *abcd* and $ABCD$ are equal:

$$ab \cdot \sin\theta_1 = AB \cdot \sin\theta_2;$$

replacing $\sin\theta$ by θ and using $ab/AB = \lambda$, we obtain the magnification factor as the ratio of the beams' widths:

$$\theta_2/\theta_1 = \lambda.$$

9.9 Problems

1. Verify that if the mixed partial derivative $\frac{\partial^2}{\partial x \partial X} P(x, X) \neq 0$, then equations (9.5) define X, Y as functions of x, y.

   Solution. According to the condition, $\frac{\partial}{\partial x} P(x, X)$ is a monotone function of X. Thus X is determined uniquely by x, y. Thus the second equation in (9.4) defines Y in terms of x, y.   ◇

2. What mapping corresponds to the generating function $P = \frac{k}{2}(X - x)^2$? How can such a mapping be realized using mechanics? Hint: See figure 9.3.

3. Find the generating function $P(x, X)$ which produces each of the maps in figure 9.1.

4. Given a linear mapping with determinant 1, find a generating function producing this mapping.

5. Consider a system with springs shown in figure 9.7. Which maps in figure 9.1 can be realized by an appropriate choice of Hooke's constants k_1, k_2, k_3?

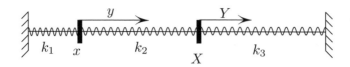

Figure 9.7. A (more) general area-preserving mapping realized via springs.

6. Which choices of Hooke's constants in figure 9.7 correspond to the scope in figure 9.6(a)?

7. (In this problem I assume familiarity with the lens formula.) Find a mechanical analog of the lens formula $\frac{1}{d_1} + \frac{1}{d_2} = \frac{1}{f}$. Hint: The lens formula expresses the fact that the rays emitted from a point refocus at another point. All such rays take the same time between the two points. For a corresponding mechanical system, potential energy is the same for different configurations, and a certain sum of forces is therefore zero.

8. Given any symmetric 2×2 matrix A, find a mechanical system whose potential energy is the quadratic form $\frac{1}{2}\langle A\mathbf{x}, \mathbf{x}\rangle$, and thus the potential force at $\mathbf{x}$ is $A\mathbf{x}$.

9. Using the mechanical realization of a symmetric matrix A from the preceding problem, prove that the eigenvalues of A are real. Hint: If the eigenvalues are not real, then there exists a perpetual motion machine. Namely, then the work done by the force $A\mathbf{x}$ around the unit circle is nonzero (no calculation is required for the proof).

10. Show that the orthogonality of the eigenvectors of a symmetric matrix A is a consequence of the non-existence of perpetual motion machine. Hint: Interpret $A\mathbf{x}$ as the force exerted by an appropriate mechanical device.

10

A BICYCLE WHEEL AND THE GAUSS-BONNET THEOREM

10.1 Introduction

This chapter tells an interesting story on how playing with a bicycle wheel can connect to a fundamental theorem from differential geometry. The internal angles in a planar triangle add up to $180°$. This fact can be restated in a more general and yet more basic way: if I walk around a closed curve in the plane, then my nose, treated as a vector, will rotate by 2π (provided that I always look straight ahead).[1]

Does the same hold for an inhabitant of a curved surface? Figure 10.1 shows a triangular path on the sphere. Two of the sides lie on meridians and one lies the equator. To a resident of the sphere the sides of the triangle appear to be straight lines.[2] A plane flying around this triangle will make three left $\pi/2$ turns between the "straight" paths, thus turning by $3\pi/2 < 2\pi$ during its round trip. In fact, for any closed path on a sphere of radius 1 the "turning angle" is, as we show later, given by

$$\theta = 2\pi - A, \tag{10.1}$$

where A is the area enclosed by the path. For the equatorial path, for example, we have $A = 2\pi$ (the area of the hemisphere), and thus $\theta = 2\pi - 2\pi = 0$, in agreement with intuition.

Expression (10.1) is a special case of the Gauss-Bonnet theorem. The theorem gives the "turning angle" θ for a closed path on any

---

[1] A precise mathematical meaning of this statement is given in section 10.3.
[2] Such "straight" lines are called the *geodesics* and defined in section 10.3.

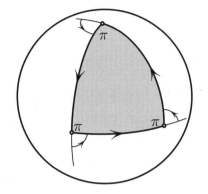

Figure 10.1. The sum of angles in a spherical triangle is $\pi - A/R^2$.

smooth surface, not necessarily a sphere. The background necessary for the theorem is described in section 10.3.

Here is the plan of the chapter. The gist of the Gauss-Bonnet theorem resides in a simply stated theorem about cones in section 10.2; this theorem is the backbone of the chapter. This theorem was literally suggested to me by my bike wheel when I was changing a punctured bike tire, via the motivation described in section 10.5. Not only was the cones theorem motivated by mechanics, but the proof is also given by mechanics (p. 136).[3] All the remaining sections are applications of the theorem on cones. These include

1. How to measure the area of a country using an inertial wheel.
2. How the precession of the wheel's axis causes rotation.
3. Background for the Gauss-Bonnet theorem (the geodesic curvature and the Gaussian curvature).
4. Gauss-Bonnet theorem as a restatement of the dual-cones theorem.

Once the statement of the dual-cones theorem is understood, the sections can be read in any order, with one obvious exception: the Gauss-Bonnet section relies on the background section preceding it.

---

[3]The purely mathematical counterpart of this proof can found in [L2].

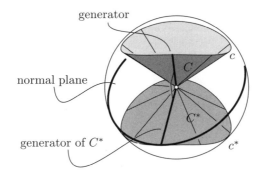

Figure 10.2. Definition of the dual cone.

10.2 The Dual-Cones Theorem

The definition of dual cones. Let C be a convex cone, as in figure 10.2. The rays making up the cone are called the *generators*. To define the dual cone, imagine the family of planes normal to the generators of C. All these planes are tangent to some invisible cone, as in figure 10.2. That latter cone is called the dual of C, denoted by C^*.

The duality turns out to be reflexive: the dual to C^* is C, that is, $(C^*)^* = C$. This is because the definition of C^* is equivalent to the following two facts:

1. The generators of C and C^* come in orthogonal pairs, as illustrated in figure 10.2.
2. The tangents to the two curves c and c^* at the corresponding points are parallel, where c and c^* are the curves of intersection of C and C^* with the unit sphere.

But these two properties are completely symmetric: neither cone is treated with preference, and thus each is the dual of the other. A detailed proof can be found in [L2].

It is intuitively clear that the sharper the cone, the "duller" the dual. The precise relationship is captured by the following simple theorem. Although simple, this theorem amounts in effect to the Gauss-Bonnet formula (as is shown in section 10.4).

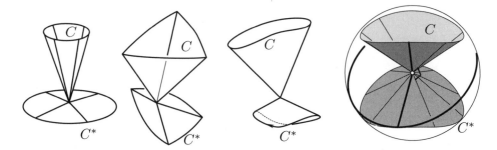

Figure 10.3. Some cones and their duals. $A(c) + L(c^*) = 2\pi$.

The dual-cones theorem. *If C is a convex cone and C^* is its dual, then*

$$A(C) + L(C^*) = 2\pi, \qquad (10.2)$$

where $A(C)$ is the solid angle of C, that is, the area on the unit sphere $\mathbb{S}^2$ enclosed by C and where $L(C^*)$ is the length of the curve $C^* \cap \mathbb{S}^2$ on the unit sphere. Note that the cones are not necessarily circular.

This purely geometrical theorem was suggested to me by mechanics, as was the proof given next.

Proof by mechanics.

The mechanical system. The generators of the cones C and C^* come in orthogonal pairs. Let us represent each generator by its unit vector, with the tail at the vertex of the cone. We thus have a bouquet of right-angle brackets held together at the origin (figure 10.4). We think of these brackets as rigid objects which can pivot freely around the origin. The ends of these brackets form the two curves c and c^* on the unit sphere $\mathbb{S}^2$. We now imagine that the sphere carries a two-dimensional gas of pressure $p = 1$, but the spherical cap enclosed by c contains a vacuum. The pressure tries to collapse the curve c. To compensate, we imagine the curve c^* to be a constant-tension spring, of tension $T = 1$, glued to the "lower" ends of the brackets. This creates a competition: each curve wants to collapse—one from

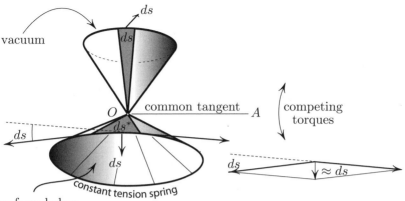

Figure 10.4. The pressure upon ds (*top*) balances the resultant of tensions upon ds^*
(*bottom*).

the pressure, the other from the tension.[4] The rigidity of the brackets
prevents simultaneous collapse.

The equilibrium. Remarkably, the mechanical system just described is in neutral equilibrium regardless of the shape of the
cone C. Postponing the proof for a moment, we note that this implies
the claim (10.2). Indeed, the potential energy then is independent of
the shape of C. But the potential energy of a vacuum bubble is $A(C)$,
while the potential energy of a constant tension spring is its length
$L(C^*)$ (see sections A.4 and A.1). Hence

$$A(C) + L(C^*) = \text{constant}.$$

By collapsing C to a point we thus expand C^* to a great circle, and
thus the constant is identified as 2π, thus proving (10.2).

   It remains to prove that the two cones are in equilibrium. Consider
two small corresponding sectors on C and on C^*. Let ds and ds^*
be the lengths of the corresponding arcs of c and c^* (figure 10.4).
The pressure force upon ds is given, to the leading term in ds, by
$p \cdot ds = ds$ and thus the torque around the direction OA parallel

---

[4]Pressure and tension can have the same effect on people.

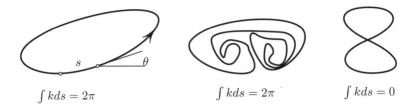

$$\int kds = 2\pi \qquad\qquad \int kds = 2\pi \qquad\qquad \int kds = 0$$

Figure 10.5. Planar curvature and its integral. During one circumnavigation of the curve the tangent turns by the angle 2π *if* the curve has no self-intersections.

to the tangent at a point on the arc equals $ds + \varepsilon$, where ε is small compared to ds: $\varepsilon/ds \to 0$ as $ds \to 0$. On the other hand, the arc ds^* is subject to two unit forces of tension; the angle between these two forces is given by the angle between the two planes tangent to the cone C^* along the two generators bounding the sector. But the *angle between these two planes equals to the angle between their normals*, that is, to ds. We conclude that the resultant force upon the arc ds^* is $2T \sin(ds/2) = ds + \varepsilon$. The torque upon ds^* has magnitude $ds + \varepsilon$. The competing torques upon ds and ds^* therefore have the same magnitude, to the leading order. Moreover, the directions of these torques are directly opposite, since the tangents at the corresponding points at c and c^* are parallel. $\diamond$

In the rest of this section we explore several consequences of the dual-cones theorem.

10.3 The Gauss-Bonnet Formula Formulation and Background

In this section I give a statement of the Gauss-Bonnet theorem; the next section shows how this theorem follows from the theorem on dual cones.

Planar curvature. The curvature of a planar curve is defined as

$$k = \frac{d\theta}{ds}$$

(figure 10.5), where s is the arclength and θ is the angle between the tangent and a fixed direction in the plane.

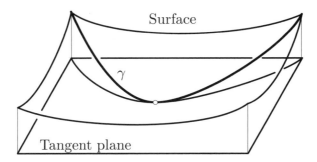

Figure 10.6. The geodesic curvature is the curvature of the curve projected on the tangent plane.

Physically, k is the angular velocity of the tangent line as the point of tangency travels along the curve with unit speed. Indeed, if the speed is 1, then s is the time, and $\frac{d\theta}{ds}$ is the rate of change of the angle with time, that is, the angular velocity.

It is intuitively clear that if I walk around a planar closed path without self-intersections, looking straight ahead all the time, then my nose will rotate by 2π:

$$\int_0^L k(s)\, ds = 2\pi; \tag{10.3}$$

formally, by the definition of k and by the fundamental theorem of calculus:

$$\int_0^L k(s)\, ds = \int_0^L \frac{d\theta}{ds} ds = \theta(L) - \theta(0) = 2\pi,$$

where the last equality expresses the fact that the tangent to a closed curve turns by 2π if the curve does not intersect itself. This fact is not as obvious as it may seem: for a "messy" curve as in figure 10.5 it may not be all that clear. A rigorous proof can be found in [CL].

The geodesic curvature. Instead of a plane we now consider a surface (figure 10.6). Imagine an ant traveling along a path γ on a surface. The tiny ant thinks that the surface is flat, and for him the curvature of γ at a point is simply the (planar) curvature of the projection of γ onto the tangent plane at the point. This "projected"

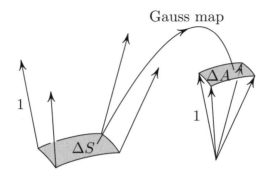

Figure 10.7. The Gaussian curvature.

curvature is called the *geodesic curvature* of γ. Integral $\int_\gamma k(s)\,ds$ has the meaning of the angle by which he rotates in one traversal of γ. Any curve whose geodesic curvature is zero will appear to the ant as a straight line. Such a curve is called a *geodesic*.

Here is a *physical interpretation: the geodesic curvature of a curve γ on a surface is the angular velocity around the normal to the surface of the tangent vector whose base point moves along γ with unit speed.*

The Gaussian curvature. To define the Gaussian curvature at a point p, we surround p by a patch of small area ΔS (figure 10.7).

Consider the "porcupine" of unit normal vectors at the patch. Let us drag each of these unit vectors over, placing their tails at a common point. The mapping thus defined assigns a point on the unit sphere to a point on the surface. This map is called the *Gauss map*, although the *mouse map* (or a *maus map*) would have been more descriptive. Let ΔA be the solid angle of the resulting cone. *Gaussian curvature* at a point p is defined as

$$K = K(p) = \lim_{\Delta S \to 0} \frac{\Delta A}{\Delta S},$$

where the limit is taken over the patches including p with diameters approaching zero. In other words, K is the Jacobian of the Gauss map. Gaussian curvature measures the "bulging" of the surface. The area ΔA is taken with a sign; for convex surfaces, such as an egg, it

Figure 10.8. A region D of a surface, bounded by a curve γ.

is positive, and thus $K > 0$ (we assume a nondegenerate case). For a saddle surface, $K < 0$. The cylinder, although curved, is not bulged and, in fact, has $K = 0$. Indeed, for the cylinder the image of a patch under a Gauss map collapses to a circular arc, whose area $\Delta A = 0$.

The Gauss-Bonnet formula. Consider a region D on a surface bounded by a curve γ, as shown in figure 10.8. The reader can think of γ as a latitude line on a sphere; however, no roundness is assumed: it could be some path on the surface of an irregularly shaped asteroid. The Gauss-Bonnet formula states that

$$\int_\gamma k\,ds + \int\!\!\int_D K\,dS = 2\pi, \qquad (10.4)$$

where k is the geodesic curvature of the curve γ, K is the Gaussian curvature of the surface, and dS is the element of the surface area.

An interpretation of the Gauss-Bonnet formula. Let us rewrite (10.4) as

$$\int_\gamma k\,ds = 2\pi - \int\!\!\int_D K\,dS.$$

The left-hand side can be interpreted as the turning angle—the angle by which a plane's heading rotates as the plane travels once around the curve γ on the surface. The theorem states that the turning angle is reduced from 2π by the amount of "bulging" $\int\!\int K\,dS$.

We interpreted (10.4), but does it have a simple explanation? In the next section I show that (10.4) comes from the dual-cones theorem. And the dual-cones theorem boils down to the fact that the angle between two planes is equal to the angle between their normals. The same therefore can be said about the Gauss-Bonnet formula. Again,

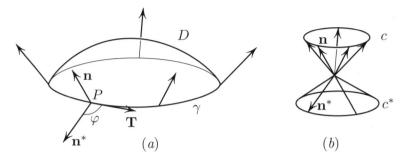

Figure 10.9. The proof of the Gauss-Bonnet theorem.

as in many other examples, something surprising and interesting (Gauss-Bonnet theorem) reduces to something surprisingly simple.

10.4 The Gauss-Bonnet Formula by Mechanics

The dual-cones theorem, which we "proved" by a simple mechanical argument, can be restated to become the Gauss-Bonnet theorem. In the final analysis this provides a mechanical proof of the latter theorem.

Proof. We start with the setting of the Gauss-Bonnet formula: a disk D on a convex surface in $\mathbb{R}^3$, bounded by a closed curve γ (figure 10.9). Consider the cone C generated by normal vectors $\mathbf{n}$ to D along γ, together with the dual cone C^*. By the dual-cones theorem

$$A(C) + L(C^*) = 2\pi. \qquad (10.5)$$

By the definition of the Gaussian curvature given previously, $K = dA/dS$, we have

$$A(C) = \int_D K \, dS.$$

It remains to show that the second term in (10.5) is the integral of the geodesic curvature. Let us drag a unit vector $\mathbf{n}^* \in c^*$ in figure 10.9(b) to the corresponding point on γ (figure 10.9(a)). Note that $L(C^*) = \int_\gamma \omega(\mathbf{n}^*)ds$, where $\omega(\mathbf{n}^*)$ is the angular velocity of the vector $\mathbf{n}^*$ around the normal direction $\mathbf{n}$, as the point P travels around γ with unit speed. Since the angle $\varphi = \angle(\mathbf{n}^*, \mathbf{T})$ is a periodic function in s, we have

$$\int_\gamma \omega(\mathbf{n}^*)\, ds = \int_\gamma \omega(\mathbf{T})\, ds.$$

This is essentially saying that if I walk once around γ, looking straight ahead at all times, then my nose will rotate by the same amount as if I kept turning my head during the trip, provided that I face in the same direction at the beginning and at the end of the trip. But $\omega(\mathbf{T}) = k$ by the definition of the geodesic curvature. This shows that $L(C^*) = \int_\gamma k\, ds$ and concludes the proof of the Gauss-Bonnet formula.   ◇

10.5 A Bicycle Wheel and the Dual Cones

The idea of dual cones was suggested by a bike wheel. As I was fixing a punctured bike tube, the following question came up: "Can one turn a bike wheel around its axle, holding the wheel only by the axle?" The wheel is initially at rest, the bearings are perfect, and the wheel is balanced, so that the wheel does not spin around its axis.

An example in figure 10.10 shows that the answer is yes. In fact, the wheel rotates through the angle given by the solid angle of the cone traced out by the axle, as we will show.[5]

One can simulate the motion in figure 10.10 with one's arm, as follows. Straighten your right arm in front of you, making the hand into a fist and holding the thumb up. The arm is the axle of an

---

[5]This rotation of the wheel is a manifestation of the so-called holonomy associated with parallel transport ([L2]). A mechanical interpretation of parallel transport is the following. Given a tangent vector, imagine that this vector is a spoke of a wheel whose disk is tangent to the surface. At no time is the wheel rotating around its axis. As the wheel is carried along a given curve on the surface, the spoke is moved in a way that is dictated by the preceding sentence. This is a mechanical interpretation of parallel transport. For a rigorous definition, see [ARN].

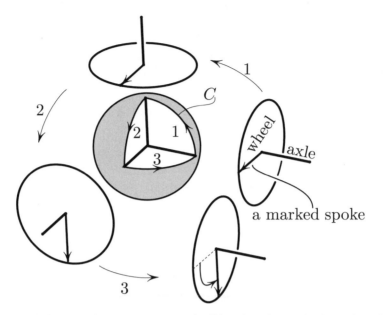

Figure 10.10. The axle traces out a cone of solid angle $\pi/2$; the wheel turns by $\pi/2$.

imaginary wheel, and the thumb is a marked spoke on that wheel, perpendicular to the arm/axle. Now do the following three motions:

1. Raise the arm above your head; the thumb will point backward. At no time are you allowed to twist your hand (as if you were turning a screwdriver); this is like the wheel not rotating around the axle.
2. Lower the arm to a horizontal position on your right. The thumb is still pointing backward.
3. Bring the arm forward in the horizontal plane to its initial position. The thumb is now pointing right. But it started up, and you didn't twist your arm!

Our discussion is based on using the inertia of the wheel to accomplish parallel transport. To accomplish parallel transport of a vector along a curve on the surface, imagine the vector as a marked spoke of a wheel. As the wheel is carried along the curve, with the plane of the wheel kept tangent to the surface, the spoke undergoes parallel transport.

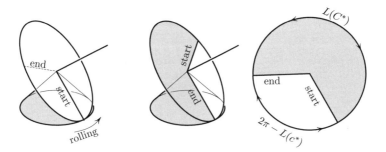

Figure 10.11. The wheel rolling on the cone C^* rotates by $2\pi - L(c^*)$.

Here is how the idea of the dual cone came from contemplating a bike wheel. Imagine the wheel as its axle sweeps a cone C (figure 10.11). The wheel executes a wobbling motion. The disk of the wheel is tangent at all times to an imaginary cone in space. This is precisely the dual cone C^*.

By how much does the wheel turn after its axle executes a conical motion? The answer comes from the following observation.

Theorem. *The plane of the wheel rolls on C^* without sliding; the spoke of contact with C^* is the instantaneous axis of the wheel's rotation.*

Proof. Consider the spoke which is in contact with C^* at a certain instant. This spoke has zero velocity: indeed, its velocity in the normal direction to the plane is zero, since the stationary cone C^* is tangent to the plane at that point. Furthermore, the velocity within the plane is zero since the wheel is not rotating around the axle thanks to perfect bearings and to the fact that the wheel started at rest. $\diamond$

Corollary. *After the axle describes a cone C, the wheel rotates through the angle $\alpha = 2\pi - L(C^*)$.*

Proof. Indeed, imagine covering C^* with wet paint. After the wheel executes one roll around C^*, a sector of the wheel will pick up the paint from C^*. Since there is no sliding, *the sector's arc has the same length as the "skirt" of the cone C^**, as shown in figure 10.11. The

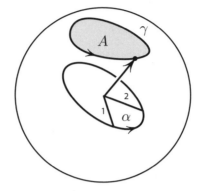

Figure 10.12. The spherimeter. The area $A = \alpha r^2$.

unpainted arc's angle is $2\pi - L(c^*)$; it is the angle by which the wheel has rotated.   ◇

By the dual-cones theorem we conclude

$$\alpha = A(C). \tag{10.6}$$

This means that *the wheel turns through the angle equal to the solid angle of the cone described by its axle.*

10.6 The Area of a Country

According to (10.6), if the axle of an initially resting wheel describes a cone C, then the wheel will turn by the angle equal to the solid angle $A(C)$ of the cone. Here is an application of this observation.

The spherimeter. Imagine a Plexiglas globe, with a needle pivoting on the globe's center; the needle can be pointed at any point on the globe. The needle also serves as the axle of a wheel, with perfect bearings.

Measuring the area inside a closed curve γ on the sphere.
We point the needle at a starting point on γ; with the wheel fixed, we mark a spoke on the wheel and remember its position. We then guide

the tip of the needle around γ, bringing it back to the starting point. By measuring the angle α through which the wheel has turned, we obtain the area A inside γ:

$$A = \alpha R^2,$$

where R is the radius of the sphere.

To be precise, the angle α is defined up to a multiple of 2π, and we have to be a bit more careful. However, if γ is confined to a hemisphere, we can choose $0 < \alpha < 2\pi$.

11

COMPLEX VARIABLES MADE SIMPLE(R)

11.1 Introduction

In this section I present some theory of complex variables, with physical insight but without rigorous proofs. No prior exposure to the theory of complex variables is assumed. One idea, used in about half of the chapter, links any complex function with an idealized fluid flow in the plane (the details are in section 11.3). With this compact idea some of the basic facts of the theory become intuitively obvious.

The first section on complex numbers requires little background. The rest of the chapter should be accessible to anyone who saw line integrals. The concepts of the divergence and curl are explained to the degree they are used.

Here are some of the highlights.

1. The Cauchy integral formula, a fundamental result, is shown to really be another conservation of mass kind of statement: if some mass of incompressible fluid is produced at a point inside a region, then the same mass must exit through the boundary of the region (section 11.5).

2. The Riemann mapping theorem, one of the most important facts of the theory, is made almost obvious by a physical interpretation. For quite a few years after I had learned the theorem, and could produce the proof upon request, I could not really explain *why* the theorem is true. The physical interpretation in section 11.7 makes the theorem very believable and can be converted into a proof.[1]

---

[1] The details of which can be found in the excellent book by Nevanlinna and Paatero [NP].

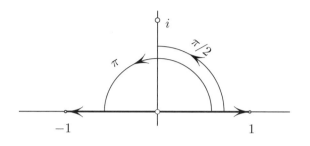

Figure 11.1. $\angle(i \cdot i) = \angle(i) + \angle(i) = \pi/2 + \pi/2 = \pi$, so that i^2 is a negative number.

3. It is a striking fact that the famous formula $1 + 1/2^2 + 1/3^2 + \cdots = \pi^2/6$ of Euler amounts to the statement that, for a certain incompressible flow of fluid in the plane with sources and a sink, the mass generated at the sources equals the mass absorbed by the sink (section 11.8).

11.2 How a Complex Number Could Have Been Invented

Multiplication revisited. When first told that $(-1) \cdot (-1) = 1$, some of my classmates and I were puzzled, thinking: "If negative is bad, how can multiplying two 'bads' be good?" Much later I realized that the fact makes perfect geomeric sense. A negative number, viewed as a vector on the line, forms the angle π with the positive x axis: $\angle(-1) = \pi$, (figure 11.1); for a positive number this angle is zero: $\angle(1) = 0$, or, we could also say, an integer multiple of 2π.

The multiplication rule, expressed geometrically, states: "*In multiplication, the angles add, while the magnitudes multiply.*"

This sheds light on the seemingly strange rule stated earlier. Indeed,

$$\angle((-1) \cdot (-1)) = \angle(-1) + \angle(-1) = \pi + \pi = 2\pi,$$

so that $(-1) \cdot (-1)$ lies along the positive x axis, that is, it is a positive number.

Introducing i. Then what is a reasonable definition of the "number" i such that $i^2 = i \cdot i = -1$? The multiplication rule stated above suggests the answer: since the angles add under multiplication, and since $\angle(-1) = \pi$, we expect $\angle(i) = \pi/2$! This leads us to define i as shown in figure 11.1, simply as the point $(0, 1)$ in the plane.[2] We have thus defined the simplest complex number, i. A general complex number is, by the definition, a point (x, y) in the plane, written for the reasons of tradition, convenience, and common sense as $z = x + iy$; the multiplication of complex numbers is defined by the rule stated above: the angles (or *arguments*) add, and the lengths (the distances to the origin) multiply. Further details can be found in any book on complex variables, for instance, [NP], [Sp].

11.3 Functions as Ideal Fluid Flows

The following simple but not-so-obvious idea forms a bridge (one of a few) from complex variables to physics:

> *Treat a function $f(z)$ as a vector field in the plane, by assigning, to each point z, the vector $\overline{f(z)}$, the complex conjugate[3] of $f(z)$.*

Why use the complex conjugate? It turns out that the vector field $\bar{f}$ has the following remarkable property:

> *If $f(z)$ is a differentiable function of a complex variable z, then its conjugate $\overline{f(z)}$, viewed as a vector field, has zero divergence[4] and zero curl (defined in the next paragraph).*

The curl (the two-dimensional case). The curl of the vector field $\mathbf{V} = \langle P(x, y), Q(x, y) \rangle$ at a point z can be defined as follows. Imagine marking the fluid with two perpendicular dashes crossing at a point z. Then curl $\mathbf{V}(z)$ is the sum of the angular velocities of these two dashes as they are carried along by the flow $\mathbf{V}$. Thus

---

[2]With equal justification we could have said $i = (0, -1)$. This is not done because of our psychological aversion to negative signs.

[3]The reason for using a conjugate value is explained shortly.

[4]Divergence is defined in section 7.3.

the curl is (twice) the averaged angular velocity of the fluid at a point. The angular velocity of the horizontal dash is easily seen to be $\frac{\partial Q}{\partial x}$ (this makes perfect sense, since this derivative measures the "shear," that is, the dependence of the vertical velocity Q on x); similarly, the angular velocity of the vertical dash is $-\frac{\partial P}{\partial y}$. Thus curl $\mathbf{V}(x, y) = \frac{\partial Q}{\partial x} - \frac{\partial P}{\partial y}$.

In intuitive terms, any two-dimensional fluid flowing in the plane with velocity $\mathbf{V}(z) = \overline{f(z)}$ at z has these two properties: (i) as an arbitrary blob of fluid is carried along, the blob's area remains constant, and (ii) the "angular velocity" of the fluid is zero at every point. Such fluid flows are called *ideal*.

It is a striking fact that all the functions[5] we studied in high school and in calculus have this fluid interpretation! Differentiability is loaded with more physical significance than one may have expected.

***Why is it that* curl $\overline{f(z)} = $ div $\bar{f}(z) = 0$?** This property of vanishing divergence and curl of $\overline{f(z)}$ is commonly known as the Cauchy–Riemann system of equations and its derivation can be found in any book on complex variables.[6] Instead of a proof, here is a geometrical explanation (convertible to a rigorous proof for a small extra fee). Since we can move the origin to any point, it suffices to deal with $z = 0$. Since f is differentiable, we have $f(z) = cz + \varepsilon$, where ε denotes higher order terms in z in the sense that $\varepsilon/z \to 0$ as $z \to 0$. Therefore, we lose no generality in assuming that f is linear: $f(z) = cz = az + ibz$, where a and b are real. The question has boiled down to showing that both $\bar{z}$ and $i\bar{z}$ have zero divergence and zero curl. Figure 11.2 makes it rather clear that this is indeed the case. For instance, div $\bar{z} = 0$ is seen from the fact that the contraction of a square centered at the origin in the y direction exactly cancels its expansion in the x direction.

Problem. *In the vortex flow corresponding to $f = i/z$ (figure 11.2), every particle of fluid rotates around the origin in a circle. How is it then possible that the curl, which measures the fluid's local angular velocity, is zero?*

[5] With the rare exceptions, such as $y = |x|$.
[6] For example, [NP], [Sp].

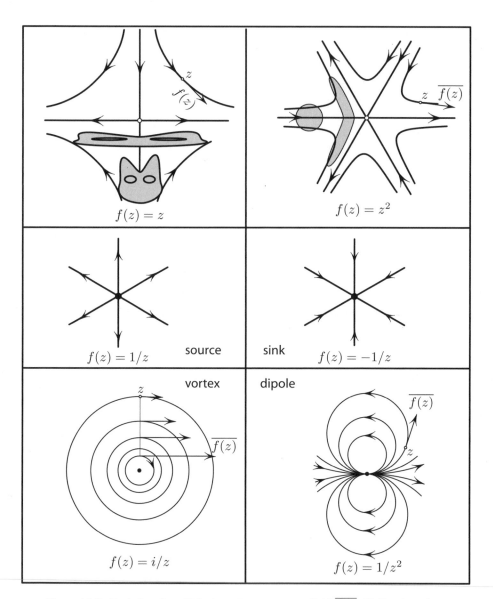

Figure 11.2. Each function $f(z)$ gives rise to a vector field $\overline{f(z)}$. If f' exists, then the corresponding flow is irrotational and incompressible. An empty circle at the origin indicates zero speed; a black dot indicates a singularity with the flow speed nearby approaching infinity.

Answer. A small dash tangent to the circle indeed rotates clockwise. However, one should not overlook the fact that the perpendicular dash rotates *counterclockwise* because the speed decreases away from the origin. The sum of the two angular velocities (the curl) turns out to be zero, as we know from the general principle, or can verify by a direct computation.

11.4 A Physical Meaning of the Complex Integral

The integral $\int_C f(z)dz$ of a complex function f along a closed curve[7] C has a nice interpretation, encoding two physical concepts in one short notation:

$$\int_C f(z)dz = \text{Circulation}_C \ \bar{f} + i \ \text{Flux}_C \ \bar{f}. \qquad (11.1)$$

Here the circulation of a vector field over C is defined as the integral of the tangential component over C, and the flux is defined as the integral of the outward normal component of V:

$$\text{Circulation}_C \ \mathbf{V} \overset{\text{def}}{=} \int_C \mathbf{V} \cdot \mathbf{T}ds, \ \ \text{Flux}_C \ \mathbf{V} \overset{\text{def}}{=} \int_C \mathbf{V} \cdot \mathbf{N}ds,$$

where $\mathbf{T}$ and $\mathbf{N}$ are the unit tangent and unit normal vectors to C, and $\cdot$ denotes the dot product.

Here is a quick proof of identity (11.1). Let $f = u + iv$, $dz = dx + idy$. Skipping some algebra, we get

$$f \ dz = \langle u, -v \rangle \cdot \langle dx, dy \rangle + i \langle u, -v \rangle \cdot \langle dy, -dx \rangle$$

(as before, $\cdot$ denotes the dot product of two vectors), or

$$f \ dz = \bar{f} \cdot \mathbf{T}ds + \bar{f} \cdot \mathbf{N}ds.$$

Integration yields (11.1).

---

[7]We do not give precise conditions that C must satisfy. For our purposes it suffices to think of C as a smooth closed curve without self-intersections.

The Cauchy-Goursat theorem. No differentiability assumptions on f were imposed yet. If we now break down and let f be differentiable, then $\mathbf{V} = \bar{f}$ becomes divergence- and curl-free.[8] We then conclude that Circulation$_C$ $\mathbf{V}$ = Flux$_C$ $\mathbf{V}$ = 0 (Green's theorem, section 7.3), and (11.1) results in the Cauchy-Goursat theorem:

If f is analytic on and inside C, then $\int_C f(z)\,dz = 0$.

As a fluid illustration of the theorem, the circulation and the flux of any of the fluid flows in figure 11.2 vanish, *provided the contour C does not enclose the singularity* (the point where f is not analytic, which is $z = 0$ in the last four examples).

11.5 The Cauchy Integral Formula via Fluid Flow

The Cauchy integral formula expresses the value of an analytic function f at any point z_0 *inside* a closed curve C in terms of the values of f *on* C:

$$f(z_0) = \frac{1}{2\pi i} \int_C \frac{f(z)}{z - z_0}\,dz. \tag{11.2}$$

Why this should be true is not immediately obvious (to most of us). However, the following *equivalent* physical statement is much more intuitive.

Physical meaning of the Cauchy integral formula. *Consider an ideal fluid flow in the plane with a source–vortex combination[9] at $z = z_0$ (figure 11.3). Then (i) the amount of fluid produced per second at the source equals the amount of fluid per second crossing the boundary C and (ii) the circulation around an infinitesimal circle surrounding z_0 equals the circulation around C.*

[8] According to the key observation on page 150.

[9] Intuitively, one can think of a thin layer of water spiraling down the drain in a sink (our picture shows the motion in reverse). Of course, the sink is not a very good example since there is friction with the bottom, the thickness of the water varies, etc.

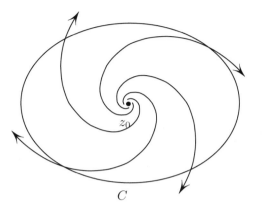

Figure 11.3. A sketch of the flow $\overline{f(z)/(z-z_0)}$ corresponding to the integrand of (11.2).

A physical significance of the right-hand side in (11.2). By (11.1),

$$\int_C \frac{f(z)}{z-z_0}\, dz = \text{Circulation} + i\, \text{Flux}, \qquad (11.3)$$

where the circulation and the flux correspond to the vector field given by the conjugate of the integrand. Let us form an idea of the nature of that flow. By Taylor's formula, $f(z) = f(z_0) + (z - z_0)g(z)$, and the integrand becomes

$$\frac{f(z)}{z-z_0} = \frac{f(z_0)}{z-z_0} + g(z) = A\frac{1}{z-z_0} + B\frac{i}{z-z_0} + g(z),$$

where $A + iB = f(z_0)$. We discovered that the flow corresponding to this function is a superposition of a source $\overline{1/(z-z_0)}$, a vortex $\overline{i/(z-z_0)}$, and a nice incompressible flow $\overline{(g(z))}$. The combination is sketched in figure 11.3.

Only the source term contributes to the flux, $2\pi A$, and only the vortex contributes to the circulation, $-2\pi B$, as figure 11.2 explains. Since $\overline{g(z)}$ is irrotational and incompressible throughout the domain,

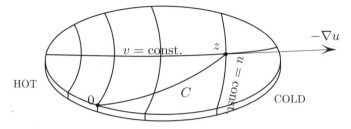

Figure 11.4. Constructing an analytic function $f = u + iv$ from a temperature distribution.

it contributes to neither. Thus (11.3) turns into

$$\int_C \frac{f(z)}{z - z_0}\, dz = -2\pi B + i2\pi A = 2\pi i(A + iB) = 2\pi i f(z_0).$$

This completes the interpretation of the Cauchy integral formula.

11.6 Heat Flow and Analytic Functions

Another remarkable physical interpretation of an analytic function, besides the one I described before, is the following. Some familiarity with gradients would be helpful in understanding this section.

The real part u of any analytic function $f(z) = u(x, y) + iv(x, y)$ can be interpreted as a stationary temperature of a plate, such as a thin metal sheet, while the imaginary part v can be interpreted as the corresponding heat flux through a curve connecting a chosen point O to (x, y). Here are the details.

Consider a planar heat-conducting negligibly thin plate D (figure 11.4) such as a thin flat copper sheet. The top and bottom surfaces of the plate are insulated, and the heat can enter or leave only through the boundary. By fixing boundary temperature and waiting an infinitely long time, we obtain a stationary temperature distribution; let $u(x, y)$ be the temperature at (x, y). The figure shows the lines of constant temperature (isotherms) $u = \text{const}$, along with a path of "heat particles"—or, more precisely, a line normal to each isotherm.

We postulate that the heat flux is $-\nabla u$. This amounts to assuming that the amount of heat crossing an infinitesimal line segment ds with the normal vector $\mathbf{N}$ is $-\nabla u \cdot \mathbf{N} ds$. In other words, we are assuming the plate's heat conductivity to be isotropic and have magnitude 1.

The law of conservation of energy imposes a special property upon u. Indeed, the net amount of heat entering any subregion is zero:

$$\int_\gamma \nabla u \cdot \mathbf{N} ds = 0, \qquad (11.4)$$

where $\mathbf{N}$ is the unit normal to γ, for any closed curve γ.

To define $v(x, y)$, let us connect an arbitrary point $z = x + iy$ with a chosen point O by a curve C, and let v be the heat flux through C:

$$v(x, y) = \int_C \nabla u \cdot \mathbf{N} ds. \qquad (11.5)$$

The integral does not depend on the curve C by property (11.4).

We now claim that $f = u + iv$ *is an analytic function.*

A sketch of the proof. Differentiating (11.5) by each of the variables and using the path independence of that integral one obtains $v_x = u_y$, $v_y = -u_x$. This is equivalent to the statement that div $\bar{f} =$ curl $\bar{f} = 0$, which is the characteristic property of an analytic function (section 11.3).

Problem. *Explain by a direct heuristic argument why the relations* $v_x = u_y$, $v_y = -u_x$ *hold.*

11.7 Riemann Mapping by Heat Flow

This discussion is not intended as a rigorous proof, but rather as a heuristic outline, meant to make the theorem intuitively obvious. Rigorous details omitted here can be found in [NP].

The background required for the discussion that follows is some familiarity with gradients.

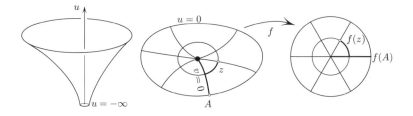

Figure 11.5. The heat enters through the boundary of D and disappears in the sink at the rate 2π cal/sec.

The Riemann mapping theorem.[10] *Let D be the open region bounded by a simple closed curve C in the complex plane, and let z_0 be a point in D. There exists an analytic function f that maps D onto the unit disk $\Delta = \{z\colon |z| < 1\}$ in a one-to-one fashion, with $f(z_0) = 0$ and $f'(z_0) > 0$.*

The physical setup. Consider a uniform heat-conducting plate D as described in section 11.6. We keep the boundary at temperature $u = 0$, and keep an infinitesimal disk at the origin so cold that 2π calories per second cross any closed curve surrounding the disk. The temperature is assumed to have stabilized and to be time independent.

The isothermal coordinates. To each point z in D we now assign two numbers: its temperature u, and the heat flux through a curve from A to z (figure 11.5), where A is a point chosen and fixed. This heat flux $v(z)$ is defined via (11.5). However, the curve Az could have several loops around the origin, and thus $v(z)$ is defined only modulo 2π, since each extra loop would pick up flux 2π. The multiple-valued function $v(z)$ can be treated of as a kind of angular variable. Now, the desired Riemann mapping is simply

$$f(z) = e^{u(z)+iv(z)}.$$

[10]This version of the theorem is weaker than the most general statement, which requires only the simple connectedness of D (see [NP]). The map f in the theorem is unique, as the Schwarz lemma implies, but we are concerned here only with its existence.

Indeed,

1. f is single-valued despite the fact that v is defined only up to 2π, since $e^{2\pi i} = 1$.
2. For z on the boundary of D we have $|f(z)| = |e^{0+iv}| = 1$.
3. $f(0) = e^{-\infty+iv} = 0$.

To achieve $f'(0) > 0$ we choose the point A (figure 11.5) whose trajectory $v = 0$ enters the origin tangentially to the x axis.

Missing rigor. Some rigorous "details" were swept under the rug; the biggest of these is the existence of the temperature distribution u. The existence of such u, called the *Green's function* of the domain, is equivalent to the existence of a solution of the so-called Dirichlet problem—a problem which is discussed in virtually all texts on partial differential equations; see, for instance, [CH].

11.8 Euler's Sum via Fluid Flow

My goal in this section is to show the picture behind Euler's formula. No proofs are given; these can be found in most texts on complex variables, for example in [Sp].

By playing with different functions $f(z)$ we can explore the resulting velocity fields $\overline{f(z)}$. One such field, corresponding to the function

$$f(z) = \frac{\cot \pi z}{2z^2}, \tag{11.6}$$

is shown in figure 11.6.

At the origin, the fluid is ejected in the x direction and absorbed in the y direction.[11] A calculation (using the Taylor series) which we omit shows that the absorption wins by $\pi^2/3$. This means that the flux through a small circle surrounding the origin is $-\pi^2/3$. On the other hand, each integer point $z = n = \pm 1, \pm 2, \ldots$, is a source of

---

[11]The speed goes to infinity near the origin. The flow pattern shown there is called the quadrupole.

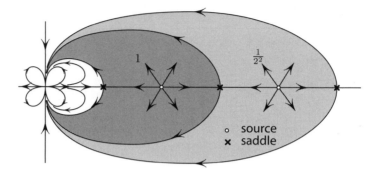

Figure 11.6. The fluid flow defined by (11.6). Each integer point n produces $\frac{1}{n^2}$ units of fluid per second. The origin absorbs $\pi^2/3$ units of fluid per second.

strength $\frac{1}{n^2}$. As the picture suggests, all the fluid coming out of the sources flows into the origin, and therefore

$$\frac{\pi^2}{3} = 2\left(1 + \frac{1}{2^2} + \frac{1}{3^2} + \cdots\right),$$

where the factor 2 comes from the fact that the sources come in symmetric pairs $\pm n$. Dividing by 2 we arrive at Euler's formula:

$$1 + \frac{1}{2^2} + \frac{1}{3^2} + \cdots = \frac{\pi^2}{6}. \tag{11.7}$$

APPENDIX

PHYSICAL BACKGROUND

This short appendix contains the physical toolbox used throughout the book.

A.1 Springs

The toolbox for our thought experiments includes two types of springs: linear springs and constant tension springs.

Zero-length springs. A zero-length spring is the one whose tension is directly proportional to its length: stretching such a string to a length x requires the force kx. Here k is a constant (called the Hooke's constant) that characterizes a particular spring. A small k means a lax spring, while a large k means a stiff spring. Note that the unstretched length of such a spring is zero.

POTENTIAL ENERGY OF A ZERO-LENGTH SPRING. By the definition, the potential energy of a spring is the work required to stretch the spring from its unstretched length (here, zero) to the length x. This work can be computed as (the average pulling force applied by my hand)·(the distance x traveled by my hand). The average force is given by $\frac{1}{2}(0 + kx) = \frac{1}{2}kx$, and thus the potential energy is

$$P(x) = \frac{1}{2}kx^2.$$

Alternatively, one can compute this work as the integral of force against the distance: $P(x) = \int_0^x (ks)ds = \frac{1}{2}kx^2$.

Constant tension springs. A spring whose tension is independent of its elongation is called a *constant tension spring*. Such a strange spring can be constructed out of a piston sliding without friction in

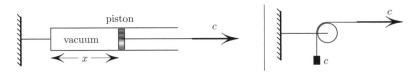

Figure A.1. Two realizations of a constant tension spring.

a cylinder enclosed at one end, with vacuum inside, as shown in figure A.1. Another realization, shown in figure A.1, consists of a weight c with a pulley.

POTENTIAL ENERGY OF A CONSTANT TENSION SPRING. By definition, the potential energy is the work required to stretch the spring from a reference length, which we take to be zero, to a given length. This work equals the force c times the distance x:

$$E = cx.$$

A.2 Soap Films

Soap films are two-dimensional analogs of constant tension springs. A soap film such as the wall of a soap bubble has an interesting property: in our idealized world, its surface tension does not change as we let our idealized film stretch or shrink. As we inflate a soap bubble its surface tension remains unchanged.

The *surface tension*, by definition, is the force required to hold together a slit of unit length. If we imagine stitching up the slit, this holding-together force is the sum of tensions of all the strings in the stitching. For soap films, the surface tension is isotropic. This means that the orientation of the slit has no effect on the surface tension. Anisotropic tensions occur in most surfaces such as skin, latex rubber films, walls of pressurized pipes, tires, and clothing materials.

The **potential energy E of a soap film** is in direct linear proportion to the film's area A:

$$E = \sigma A, \tag{A.1}$$

where σ is the surface tension. This relationship makes soap films useful in solving area-minimization problems. To prove the

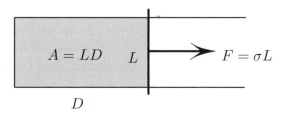

Figure A.2. Potential energy of a soap film with constant surface tension σ is σA.

relationship (A.1), consider a frame in figure A.2; the rod slides along the frame, pulling the soap film behind. If L is the length of the rod in contact with the film, then the force required to move it is $F = \sigma L$.

(Equivalently, one can think of a cylinder with a piston in two dimensions, with a vacuum inside and with the pressure σ outside.) By the definition, the energy of a configuration equals the work required to pull the rod from a reference position, which we take to be the zero-area position. To pull the rod (or a piston) a distance D takes work

$$E = \sigma L \cdot D = \sigma A,$$

proving the claim.

Problem. *Frozen pipes always burst lengthwise. Why?*

Answer. Longitudinal tension in a pressurized cylinder turns out to be twice the equatorial tension, as we now show. Let us compare the forces required to hold together a longitudinal slit versus a transversal slit (figure A.3).

Let p be the pressure inside the pipe; we assume it to be isotropic. (So, to be completely honest, we are really explaining why a pipe with compressed fluid, rather than ice, will burst lengthwise.) If we slice the pipe along the equator (figure A.3) it will take force $p \cdot \pi r^2$ to hold the cut together against the pressure p over the area πr^2. This force is held by the length $2\pi r$ of the cut, so that the force per unit

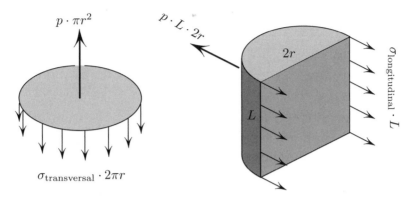

Figure A.3. Force balance along the pipe and across the pipe.

length is

$$\sigma_{\text{transversal}} = \frac{p \cdot \pi r^2}{2\pi r} = \frac{1}{2}pr.$$

If we slice the pipe along two generators each of length L (figure A.3), a force of $p \cdot (L \cdot 2r)$ will be required to hold the slit together against the pressure over the area of the rectangular section $L \cdot 2r$. Per unit length of the slit this gives

$$\sigma_{\text{longitudinal}} = \frac{p \cdot (L \cdot 2r)}{2L} = pr.$$

We found that $\sigma_{\text{longitudinal}} = 2\sigma_{\text{transversal}}$. The pipes burst lengthwise because the tension along the cylinder's generators is twice the tension along its equators.

A.3 Compressed Gas

For our purposes in this book we consider a simplified model of gas whose *pressure* p *does not change as the container changes its size:* $p = $ constant. In that sense the fictitious gas is an exact analog of soap film, except that the tension σ is negative, the opposite of pressure: $p = -\sigma, \sigma < 0$.

Potential energy of compressed gas. Consider a planar region D filled with a two-dimensional gas. Recall that the pressure is assumed to remain constant as the region changes its area. The potential energy of this system is $-pA$. The proof is a verbatim repetition of the corresponding argument for the surface tension.

Similarly, the potential energy of a three-dimensional region of volume V, filled with gas at pressure p, is $-pV$.

A.4 Vacuum

Imagine a region in the plane, with vacuum inside and with compressed gas of pressure p outside. Work pA is needed to open such a bubble of vacuum, where A is the area of the region. Indeed, consider a piston of length L in a cylinder, as suggested by figure A.2. Force $F = pL$ is needed to move the piston; to move this piston a distance D it takes work $FD = pLD = pA$. This proves that the kinetic energy of the bubble is

$$E = pA,$$

at least for a rectangular bubble. Any other shape can be approximated with any precision by small rectangles, and the above result still holds.

A two-dimensional bubble embedded in the ambient pressure p is a mathematical equivalent of a soap film with surface tension $\sigma = p$.

The same arguments apply in three dimensions, with a similar result: the potential energy of a bubble of vacuum of volume V in the ambient pressure p, that is, the work required to open such a bubble from zero, is in direct proportion to the volume:

$$E = pV. \tag{A.2}$$

A.5 Torque

Definition. Consider a force $\mathbf{F}$ applied at a point A, and let O be a chosen point, called the pivot point. The *torque* of the force $\mathbf{F}$ with respect to the pivot O is the cross product $\mathbf{T} = \mathbf{L} \times \mathbf{F}$, where $\mathbf{L} = \overline{OA}$ is the position of A relative to O, called the *lever*. The torque is also referred to as the *moment* of the force $\mathbf{F}$ with respect to the pivot point O.

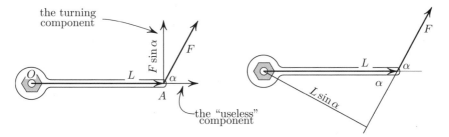

Figure A.4. Magnitude of torque $= (F \sin \alpha)L = F(L \sin \alpha) = \|\mathbf{F} \times \mathbf{L}\|$. The direction of torque is out of the paper toward the reader. It is the direction of the motion of the nut along the bolt with a right-handed thread.

This definition fits perfectly with our common sense of the "turning intensity."[1] Imagine trying to turn a stuck nut with a wrench. Only the component of $\mathbf{F}$ perpendicular to the handle $\mathbf{L}$ is useful. The magnitude of this component is $F \sin \alpha$. Not only the force, however, but length L as well affects the intensity of turning. In fact, what really matters for getting the nut unstuck is the product $(F \sin \alpha)L$. This explains the reasonableness of the definition of the magnitude of the torque as $T = \|\mathbf{L} \times \mathbf{F}\| = L(F \sin \alpha)$. But there is also a natural axis—that of the bolt on which the nut sits, perpendicular to both $\mathbf{L}$ and $\mathbf{F}$. There is also a natural direction along this axis: the direction in which the nut will move if it gets unstuck; the thread on the bolt is assumed to be right-handed by accepted convention.

A.6 The Equilibrium of a Rigid Body

For the purposes of our discussion, a rigid body is a collection of a finite number of point masses m_k with fixed distances from each other.

---

[1] We avoid the term "turning **force**" because what matters is not the force but rather the product of the force and the lever. Even a tiny force can turn a tough nut with a large enough lever.

Applying Newton's first law to a rigid body yields the following:

Theorem 1. *If a body is in equilibrium, then the sum of all forces and the sum of all torques (relative to some pivot) acting upon the body is zero.*

The choice of pivot is immaterial if the sum of forces is $\sum \mathbf{F}_k = \mathbf{0}$ since then the sum of torques does not depend on the choice of the pivot. Indeed, let $\mathbf{r}_k$ be the positions of the point masses relative to the origin, and let $\mathbf{a}$ be the position vector of the pivot. Then the sum of torques relative to $\mathbf{a}$ is the same as the sum of torques relative to the origin:

$$\mathbf{T}_A = \sum \mathbf{F}_k \times (\mathbf{r}_k - \mathbf{a}) = \sum \mathbf{F}_k \times \mathbf{r}_k - \left(\sum \mathbf{F}_k \right) \times \mathbf{a}$$
$$= \sum \mathbf{F}_k \times \mathbf{r}_k = \mathbf{T}_O,$$

as claimed. The following simple remark will be used in several geometrical problems on minima.

Lemma on three concurrent forces. *If the body under the influence of three forces is in equilibrium, then the lines of these forces are concurrent (that is, the lines share a common point).*

Proof. The sum of torques is zero relative to any pivot point, according to the above remark. Let us choose for a pivot the point of intersection[2] P of the lines of $\mathbf{F}_1$ and $\mathbf{F}_2$ (figure A.5). With this choice, the torques of $\mathbf{F}_1$ and $\mathbf{F}_2$ relative to P vanish. Since the sum of all torques is zero, the torque of $\mathbf{F}_3$ relative to P vanishes as well. But this implies that the line of $\mathbf{F}_3$ passes through P, proving that the three lines are concurrent. $\diamond$

A.7 Angular Momentum

Angular momentum is the rotational analog of the linear momentum. For a point mass m moving with speed v in a circle of radius r in

---

[2]It can happen that the two lines are parallel, in which case P is at infinity. The proof still goes through if we adjust our language, by referring to three parallel lines as concurrent (at infinity).

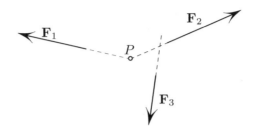

Figure A.5. The net torque relative to P is nonzero if the lines of forces are not concurrent.

the plane, the angular momentum is defined as $mv \cdot r$. But what if the mass travels in some other way, say in a straight line? In that case, the definition is the same, except that instead of v one takes the component of velocity perpendicular to the particle's position vector. Here is a precise definition.

Definition. Let $\mathbf{r} = \mathbf{r}(t)$ be the position vector of a point P of mass m relative to a point O in space. The angular momentum $\mathbf{L}$ of P with respect to O is defined as the cross product of the vectors of position and the linear momentum:

$$\mathbf{L} = \mathbf{r} \times m\dot{\mathbf{r}},$$

where $\dot{\mathbf{r}}$ stands for $d\mathbf{r}/dt$. By differentiating, we obtain the rotational analog of Newton's law $F = ma$:

$$\frac{d}{dt}\mathbf{L} = \mathbf{r} \times m\ddot{\mathbf{r}}. \qquad (A.3)$$

According to Newton, $m\ddot{\mathbf{r}} = \mathbf{F}$ is the resultant force acting on the mass. The right-hand side $\mathbf{r} \times \mathbf{F} = \mathbf{T}$ is the torque of that force upon the mass relative to O. Equation (A.3) then becomes

$$\frac{d}{dt}\mathbf{L} = \mathbf{T}, \qquad (A.4)$$

the rotational analog of Newton's second law $\frac{d}{dt}\mathbf{p} = \mathbf{F}$, where $\mathbf{p} = m\mathbf{v}$ is the linear momentum.

For a collection of masses, the angular momentum is defined as the sum of the angular momenta of the constituent parts.

Conservation of angular momentum. *If the sum of all exernal torques upon any system of masses is zero, then the angular momentum of the system is constant.*

This is a consequence of (A.4) combined with Newton's third law.

A.8 The Center of Mass

The center of mass of a rigid body is, by the definition, the point on which the body is in equilibrium in any orientation relative to a constant gravitational field. As mentioned before, we treat rigid bodies as collections of point masses with mutual distances fixed. The discussion that follows carries almost verbatim for continuous mass distributions except that the sums $\Sigma a_k m_k$ have to be replaced by integrals $\int a(x) dm$.

Theorem 2. *The center of mass $\bar{\mathbf{r}}$ of a system of point masses m_k with position vectors $\mathbf{r}_k$ is given by the weighted average of these positions:*

$$\bar{\mathbf{r}} = \frac{\sum m_k \mathbf{r}_k}{m} \equiv \sum \mu_k \mathbf{r}_k, \quad where \quad m = \sum m_k \quad and \quad \mu_k = \frac{m_k}{m}.$$
(A.5)

Note that μ_k is the proportion of the mass m_k to the total mass, so indeed the average is weighted according to the mass of each point.

Proof. By the definition of the balance point $\bar{\mathbf{r}}$, the sum of torques of gravitational forces upon each mass with respect to this point is zero:

$$\sum (\mathbf{r}_k - \bar{\mathbf{r}}) \times (m_k \mathbf{g}) = 0,$$

where $\mathbf{g}$ is the direction of gravitational acceleration. This equation must hold for any vector $\mathbf{g}$, reflecting the fact that the body must balance in any orientation; rather than turning around the body, we imagine turning the direction of gravity. That is, we choose the coordinate system attached to the masses. Multiplying out in the last equation and factoring $\mathbf{g}$ out of the sum, we obtain

$$\left(\sum m_k \mathbf{r}_k - m\bar{\mathbf{r}}\right) \times \mathbf{g} = \mathbf{0}, \quad \text{where} \quad m = \sum m_k.$$

Since $\mathbf{g}$ is an arbitrary vector, we must have $\sum m_k \mathbf{r}_k - m\bar{\mathbf{r}} = \mathbf{0}$, which gives (A.5).   ◇

A.9 The Moment of Inertia

For our purposes, we need to discuss only the two-dimensional case of planar bodies rotating within their plane. The moment of inertia is a measure of rotational inertia just like the mass is the measure of translational inertia. For a point mass m at the distance r to a point O the moment of inertia relative to O is defined to be mr^2 for the reason to be explained shortly. For a collection of masses, the moment of inertia relative to O is defined to be the sum (or an integral if the mass distribution is continuous) of moments of inertia of the constituent parts. The above definition must be explained for it to be of any use. We start with a point mass m at a distance r from a point O. Think of m as attached by a massless stick of length r to the origin, around which it can pivot. Let us apply a force F to the point, in the direction normal to the stick, to give it angular acceleration around O. We want to quantify the rotational inertia of our mass on the stick.

The translational inertia is quantified by the mass $m = \frac{F}{a}$, the ratio of the force to acceleration. Guided by this analogy, we define the moment of inertia I as the ratio of the "angular force," that is, the torque, to the angular acceleration:

$$I \stackrel{\text{def}}{=} \frac{\text{torque}}{\text{angular acceleration}} = \frac{T}{\ddot{\theta}} = \frac{rF}{\ddot{\theta}}.$$

After substituting $F = ma = mr\ddot{\theta}$ into the last expression, the term $\ddot{\theta}$ cancels and we obtain

$$I = mr^2.$$

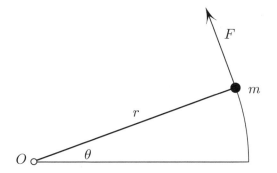

Figure A.6. The moment of inertia of a point mass: torque/angular acceleration $= mr^2$.

This explains why the definition is useful. The generalization to n point masses is straightforward: let r_k be distance from the point O to the kth mass $m_k, k = 1, \ldots, n$. The moment of inertia with respect to the point O of the collection of the masses is defined as

$$I = \sum m_k r_k^2.$$

Theorem 3. (parallel axes theorem) *Let C be the center of mass of the collection of masses m_k, and let O be an arbitrary point. Then*

$$I_O = I_C + m|OC|^2,$$

where I_C and I_O are the moments of inertia relative to the points C and O. In particular, $I_O \geq I_C$. That is, the moment of inertia around the center of mass is minimal.

Proof. Let $\mathbf{r}_k$ be the vector from O to the mass m_k. Let $\mathbf{c} = \overline{OC}$ be the vector from O to the center of mass C; we have $\mathbf{c} = \sum \mu_k \mathbf{r}_k$. Now

$$I_O = \sum m_k \mathbf{r}_k^2 = \sum m_k(\mathbf{r}_k - \mathbf{c} + \mathbf{c})^2$$
$$= \sum \left(m_k(\mathbf{r}_k - \mathbf{c})^2 + 2m_k(\mathbf{r}_k - \mathbf{c}) \cdot \mathbf{c} + m_k \mathbf{c}^2 \right).$$

The sum of the middle terms vanishes by the definition of $\mathbf{c}$, leaving the remaining terms I_C and $m|OC|^2$.   ◇

A.10 Current

Consider an elecric current, that is, the "gas" of electrons flowing through a wire. Pick any transversal cross section of the wire. Consider the amount of charge $q(t)$ crossing this section from some initial moment up to time t. Then

$$I = \frac{dq}{dt} = \dot{q}(t)$$

gives us the instantaneous rate at which charge passes through the section per second. This rate is called the **current**, or the **amperage**. The current is the same for every cross section of a wire, since (i) the number of electrons between two fixed sections is approximately equal to the *constant* number of ions in the metal to maintain zero net charge of the wire, and (ii) the electrons do not leave through the walls of the wire.

The electric current is the exact analog of the **flux** of water through the pipe. This flux can be measured in gallons per second, just like the current is measured in coulombs (units of charge) per second.

A.11 Voltage

Consider an electrostatic field created by one or more charges. Fix a reference point O in space, and let A be any other point. Define the voltage $V(A)$ at A as the potential energy, relative to O, of the unit charge at A. In other words, $V(A)$ is defined as the work required to bring the unit charge from O to A:

$$V(A) = - \int_{O}^{A} \mathbf{E} \cdot \mathbf{T} ds.$$

Here $\mathbf{E}$ is the electrostatic force acting upon a unit charge and $\mathbf{T}$ is the unit tangent vector to the path[3] OA. Note that the minus sign is

---

[3]This integral does not depend on the choice of the path from O to A. This independence follows from the law of conservation of energy. Indeed, had the integral along two paths with the same ends been different, the integral over a cycle obtained by concatenating the two paths

due to the fact that *we* have to apply the compensating force $-\mathbf{E}$ to move the charge against the electrostatic field.

Taking the gradient of both sides in the last equation, we obtain the force $\mathbf{E}$ acting upon the unit charge via the gradient of V:

$$\mathbf{E} = -\nabla V.$$

One can take this relationship as the (implicit) definition of V, equivalent to the one we gave above. It would be shorter, but less clear. Note that not every vector field is a gradient of a scalar function.[4] The field given by a gradient of a function V is called a potential field, and the function V is called the potential.

This is disussed in more detail in most vector calculus texts, for example, [St].

A.12 Kirchhoff's Laws

Kirchhoff's laws apply to any electric circuit, which consists of capacitors, resistors, inductances, batteries, diodes, and so on.

Kirchhoff's first law is a restatement, for a special case, of the fact that in an electrostatic field, zero work is done in carrying a charge around a closed path.

Kirchhoff's first law. The sum of voltage drops along any closed path of a circuit is zero.

Kirchhoff's second law. The sum of currents entering a node equals the sum of currents exiting a node.

The second law is a consequence of the conservation of electric charge: whatever enters the node must leave it.

---

would be nonzero. That is, we would then have the field do nonzero work in moving a charge along a closed path, contradicting the conservation of energy.

[4]The velocity field of a rotating disk is an example of a vector field which is not a gradient of any function. Shear wind velocity field above the land or sea is another: such velocity $\mathbf{V}$ is horizontal, with speed increasing with height. Albatrosses use the nongradient nature of this field, that is, the nonvanishing of $\oint \mathbf{V} \cdot \mathbf{T} ds$, to extract energy from the wind. These birds can soar indefinitely, just steering but not flapping, without updrafts!

A.13 Resistance and Ohm's Law

Consider again a steady current running through a piece AB of a wire. The wire is not a perfect conductor—the electrons bounce into ions on their way, losing their energy in collisions; they regain energy from the pull of the electric field applied along the wire. This is like sand falling through a series of horizontal meshes. The grains bump against the mesh, slow down, fall through, accelerate due to gravity, hit the next mesh, and so on. It is natural to expect that if the gravity is increased, the flux of sand would increase. The same happens with the electric current. The more voltage we apply, the stronger the current I becomes; in fact, experiments show that the relationship is linear. This is Ohm's law:[5]

$$\frac{V}{I} = R = \text{constant}. \qquad (A.6)$$

The coefficient R is called the **resistance**. This term is in agreement with common sense: a large resistance indicates that a large voltage is needed to achieve the same current. According to equation (A.6), R is the voltage required to produce one unit of current.

A.14 Resistors in Parallel

What is the resistance R of two resistors R_1 and R_2 connected in parallel, as shown in figure A.7? The answer is very simple:

$$\frac{1}{R} = \frac{1}{R_1} + \frac{1}{R_2}, \quad \text{or} \quad R = \frac{R_1 R_1}{R_1 + R_2}.$$

This makes perfect intuitive sense: indeed, in connecting the resistors in parallel we enhance the conductive ability by giving the current more paths to travel. The formula says that, in fact, the *conductances*—the reciprocals of the resistances—add. Here is the proof.

---

[5]This law is true only approximately, but with good enough precision for most conductors at room temperature. Nature does not strictly enforce this law.

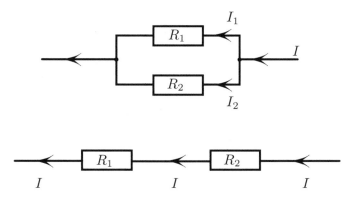

Figure A.7. For parallel connections, the current splits between the two resistors, while the voltage drop is common. For the series connection, the current is common, while the total voltage drop is split between the two resistors.

The first thing to observe is that the current splits, according to Kirchhoff's law:

$$I = I_1 + I_2. \tag{A.7}$$

Also, the voltage drop V across both resistors is obviously the same, and Ohm's law gives

$$I = V/R, \; I_1 = V/R_1, \; I_2 = V/R_2.$$

Substituting the last line into (A.7) proves the parallel resistors formula.

The resistance of $n \geq 2$ parallel resistances $R_1, \ldots, R_n$ is given by the same rule as above: $\frac{1}{R} = \frac{1}{R_1} + \cdots + \frac{1}{R_n}$.

A.15 Resistors in Series

When two resistors are connected in series (figure A.7), their resistances add:

$$R = R_1 + R_2. \tag{A.8}$$

Indeed, the voltage drop across the combination is the sum of voltage drops on each of the resistors:

$$V = V_1 + V_2. \tag{A.9}$$

Now the currents through both resistors are the same by Kirchhoff's law. Ohm's law gives

$$V = IR, \ V_1 = IR_1, \ V_2 = IR_2,$$

which, when substituted into (A.9), results in (A.8).

A.16 Power Dissipated in a Resistor

The current I passing through a resistor causes energy to be lost in the resistor, in the form of heat. The power dissipated on the resistor, that is, the amount of heat per unit of time, is given by

$$P = IV,$$

where V is the voltage drop across the resistor. The proof of this fact is essentially a restatement of the definition of V and I. Here are the details. Consider first just one electron traveling from one end of the resistor to the other. Like a pinball, the electron "falls" under the pull of the electric field, bumping into ions and giving them part of its kinetic energy, making them vibrate and thus producing heat. On average, the electrons exit the resistor no faster than they entered. Therefore, the electrons give up all of the kinetic energy they gain from the electric "pull" as heat. By the definition of the voltage, this energy equals $(\Delta q)V$, where Δq is the charge of the electrons passing through the resistor. Per unit time, we have

$$\frac{(\Delta q)V}{\Delta t} = \frac{\Delta q}{\Delta t}V = IV.$$

A.17 Capacitors and Capacitance

A capacitor is a device consisting of two conducting plates separated by a thin insulating layer. Let us connect a battery to the two plates of the capacitor. The battery would "suck" some electrons out of one plate and pump them into the other. The combined charge q of these transferred electrons is proportional to the voltage V of the battery:

$$\frac{q}{V} = C = \text{constant.} \tag{A.10}$$

The coefficient C is called the **capacity**. It can be thought of as the amount of charge that the capacitor can absorb while increasing its voltage by 1; the term "capacity" is therefore justified.

The thinner the insulating layer, the higher the capacity turns out to be. Here is a simple explanation. One of the plates has an excess of electrons, which repel each other and try to rush back out through the wire, through the battery, and into the other plate, just like a compressed gas in a vessel; the battery, like a pump, keeps them in. How does proximity of another plate affect this desire to escape? The other plate has a shortage of electrons, which is to say it is charged positively. This attracts the electrons, the more so the closer the plates are. If the plates are very close together, then less of a voltage is required to keep the electrons from escaping. That is, the capacity is then greater.

The capacitor is an electric analog of the spring, and (A.10) is the analog of Hooke's law $F = kx$ for a zero-length spring. The two laws $V = C^{-1}q$ to $F = kx$ are analogous term-by-term. In particular, C^{-1} is the "stiffness" of the capacitor, and is analogous to Hooke's constant k, which measures the stiffness of the spring. It would be reasonable to refer to $\frac{1}{k}$ as the capacitance of a spring.

A.18 The Inductance: Inertia of the Current

"Inductance" in water flow. Consider water flowing in a straight pipe. We consider an idealized world with no viscosity or turbulence. Let p be the pressure difference between two transversal sections A and B of the pipe, and let f be the flux, that is, the mass of fluid per unit time crossing a transversal section of the pipe.[6] Newton's second

---

[6]The section doesn't matter—f is the same for them all since water is incompressible.

law ($F = ma$) applied to the body of fluid that is between A and B at some instant gives

$$p = \mu \frac{d}{dt} f, \qquad\qquad (A.11)$$

where μ characterizes inertia in reacting to the pressure difference. The last equation comes from Newton's law as follows. The force upon the fluid cylinder between the sections A and B is given by $F = pS$, where S is the cross-sectional area of the pipe. Newton's second law applied to this fluid cylinder gives

$$\underbrace{pS}_{F} = \underbrace{\rho Sh}_{m} \underbrace{\dot{v}}_{a}, \qquad\qquad (A.12)$$

where v is the speed of the fluid and h is the distance between A and B. But $Sv = f$ is the flux (volume per second), and (A.12) becomes $pS = \rho h \frac{d}{dt} f$. Dividing by S we obtain (A.11) with $\mu = \rho h / S$.

Electrical inductance. Electric current has inertia as well. The mechanical inertia of moving charges plays, of course, a negligible role; the effect is rather electromagnetic and, as it turns out, explained by special relativity. We only state the mathematical fact and do not describe what happens physically when the current changes.

Any change of current through a coil requires a voltage difference between the ends of the coil; this is the manifestation of inductance. It turns out that the relationship between the voltage and the rate of change of the current is linear:

$$V = L\dot{i}. \qquad\qquad (A.13)$$

The coefficient L is called the **inductance**. According to the formula, the inductance L is the voltage required to increase the current by 1 amp per second.

A shocking example. Inductance can be used to get a painful electric shock from a small battery such as a 1.5-V AA type. Consider the circuit in figure A.8. With the switch closed, most of the current goes through the switch, and only a tiny imperceptible

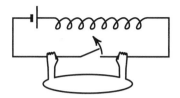

Figure A.8. As the switch is opened, the current must continue by "inertia" and is thus driven through the body.

amount trickles through me. Now what happens when the switch is opened? The current has inertia and does not like to stop suddenly; it will continue flowing by "inertia" for a (short) time through the only available path: me.[7] To put it differently, opening the switch causes a large $\dot{I}$ which, via (A.13), causes a large and painful voltage V.

Like most electrical phenomena, this one has a mechanical analog, also painful—that of stopping a hammer aimed at a nail with a fingernail. A large deceleration $\dot{v} = a$ causes a large force $F = m\dot{v}$, which in turn causes pain.

Some electrical erector sets use a ringing bell for the switch. As the oscillating hammer hits the bell, a switch repeatedly opens and closes and the user receives what feels a continuous shock.

This completes our painful lesson on inertia.

A.19 An Electrical-Plumbing Analogy

All of the above concepts—V, I, q, R, C, and L—have a simple analog in plumbing (such as a house plumbing!). Figure A.9 summarizes the analogy.

---

[7]It is interesting to note that as the switch breaks the contact, the current can still continue to flow through the air as a spark. A related effect called the water hammer is familiar to some homeowners who hear banging pipes when the washer shuts off: the water flow is stopped by a valve that shuts very quickly, causing a great spike in pressure from the water, which wants to keep flowing.

ELECTRICAL   A PLUMBING ANALOG

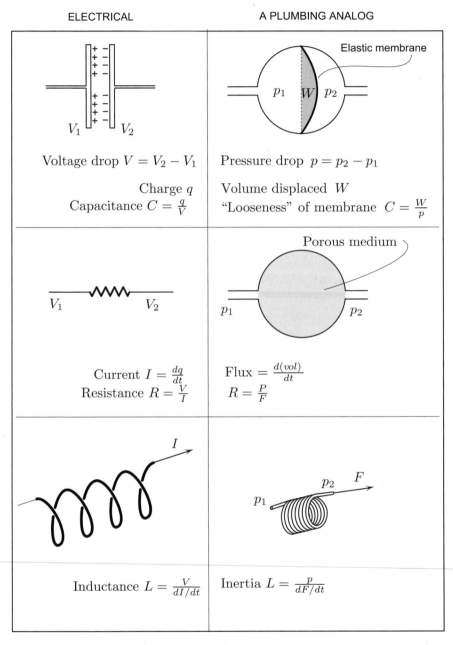

Voltage drop $V = V_2 - V_1$ Pressure drop $p = p_2 - p_1$

Charge q Volume displaced W
Capacitance $C = \frac{q}{V}$ "Looseness" of membrane $C = \frac{W}{p}$

Current $I = \frac{dq}{dt}$ Flux $= \frac{d(vol)}{dt}$
Resistance $R = \frac{V}{I}$ $R = \frac{P}{F}$

Inductance $L = \frac{V}{dI/dt}$ Inertia $L = \frac{p}{dF/dt}$

Figure A.9. The electrical-plumbing analogy.

A.20 Problems

1. Consider a point charge q placed at the origin. Using the definition of voltage in section A.11, find the voltage at any point in space.

The following three problems require no knowledge beyond the understanding of the definition of voltage.

2. Using the definition of voltage in section A.11, find the speed of the electrons hitting the screen in a TV monitor, given the voltage difference V between the cathode (from which the elecrons emanate into the vacuum inside the tube) and the screen, the mass m, and the charge q of an electron. Hint: The potential energy of the electrons at the beginning of the flight is converted to the kinetic energy at the end. V is the potential energy difference for a **unit** charge.

3. A capacitor of capacitance C is charged to voltage V. Find the potential energy of the capacitor.

   Solution. This problem can be done with integrals, but here is a way to avoid them. Imagine that we charge the capacitor from zero voltage to V by pumping a constant current. The average voltage during this process is $\frac{1}{2}V$ (since the charge, and hence the voltage, grows linearly with time). We can thus pretend that even as we carry charge piece-by-piece from one plate to the other, the voltage remains the same: $V/2$. Thus the total work to charge the capacitor is $W = q \cdot V/2 = CV \cdot V/2 = CV^2/2$.

4. The following statement came from an online tutorial: "From the definition of voltage as the energy per unit charge, one might expect that the energy stored on this ideal capacitor would be just qV. That is, all the work done on the charge in moving it from one plate to the other would appear as energy stored. But in fact, the expression above ($qV/2$) shows that just half of that work appears as energy stored in the capacitor. For a finite resistance, one can show that half of the energy supplied by the battery for the charging of the capacitor is dissipated as heat in the resistor, regardless of the size of the resistor."

   Can you prove or disprove this statement?

5. Find the power required to run a light bulb with voltage V and resistance R.

Solution. The voltage difference between the two ends of a bulb filament is V. This means, by the definition of voltage, that the charge q of electrons loses energy qV in passage through the filament.[8] The power expended is the energy per unit time: $P = qV/t = IV$, where $I = q/t$ is the current (by the definition of the current). By Ohm's law $I = V/R$, and

$$P = IV = \frac{V^2}{R}.$$

This formula explains why one gets an explosion when shorting an outlet: a small R makes for a large P.

---

[8]The electrons "fall" down the electrostatic potential along the filament. The energy they lose in collisions with ions goes into agitating the ions in the filament and becomes heat and light.

BIBLIOGRAPHY

[ARC]   Archimedes. *Geometrical Solutions Derived from Mechanics*, trans. J. L. Heiberg. Chicago: Open Court Publishing Company, 1909. PDF file available in http://books.google.com/books?id= suYGAAAAYAAJ.

[ARC1]   Archimedes' Palimpsest. http://www.archimedespalimpsest.org/.

[ARN]   V. I. Arnold. *Mathematical Methods of Classical Mechanics*, trans. K. Vogtmann and A. Weinstein. New York: Springer-Verlag, 1978.

[BB]   M. B. Balk and V. G. Boltyanskii. *Geometriya mass* (Russian) [The geometry of masses]. Bibliotechka Kvant [Library Kvant], 61. Moscow: Nauka, 1987.

[CG]   H.S.M. Coxeter and S. L. Greitzer. *Geometry Revisited*. Washington, DC: Mathematical Association of America, 1967.

[CH]   R. Courant and D. Hilbert. *Methods of Mathematical Physics*. Vol. II. *Partial Differential Equations*. Reprint of the 1962 original. Wiley Classics Library. New York: Wiley-Interscience, 1989.

[CL]   E. Coddington and N. Levinson. *Theory of Ordinary Differential Equations*. New York: McGraw-Hill, 1955.

[D]   M. M. Day. Polygons circumscribed about closed convex curves. *Trans. Am. Math. Soc.* 62 (1947), pp. 315–319.

[DO]   M. DoCarmo. *Differential Geometry of Curves and Surfaces*. Englewood Cliffs, NJ: Prentice-Hall, 1976.

[DS]   P. G. Doyle and J. L. Snell. *Random Walks and Electric Networks*. Washington, DC: Mathematical Association of America, 1984. See pages 65–69 for further details and references.

[Fe]   R. P. Feynman. *QED*. Princeton, NJ: Princeton University Press, 1985.

[Fo]   R. L. Foote. Geometry of the Prytz planimeter. *Rep. Math. Phys.* 42(1–2) (1998), pp. 249–271.

[GF]   I. M. Gelfand and S. V. Fomin. *Calculus of Variations*. Engle-
         wood Cliffs, NJ: Prentice-Hall, 1963.

[HZ]   H. Hofer and E. Zehnder. *Symplectic Invariants and
         Hamiltonian Dynamics*. Birkhäuser Advanced Texts/Basler
         Lehrbucher. Basel: Birkhäuser Verlag, 1994.

[K]   B. Yu. Kogan. *The Applications of Mechanics to Geometry*.
         Chicago: University of Chicago Press, 1974.

[L1]   ———. Minimal perimeter triangles. *Am. Math. Monthly* 109
         (2002), pp. 890–899.

[L2]   M. Levi. A "bicycle wheel" proof of the Gauss-Bonnet theorem,
         dual cones and some mechanical manifestations of the Berry
         phase. *Expo. Math.* 12 (1994), pp. 145–164.

[LS]   Yu. I. Lyubich and L. A. Shor. *The Kinematic Method in Geo-
         metrical Problems*, trans. V. Shokurov. Moscow: Mir Publishers,
         1980.

[LW]   M. Levi and W. Weckesser. Non-holonomic effects in averaging.
         Erg. Th. & Dynam. Sys. 22 (2002), pp. 1497–1506.

[M]   J. Milnor. *Morse Theory*. Annals of Mathematics Studies,
         No. 51. Princeton, NJ: Princeton University Press, 1963.

[NP]   R. Nevanlinna and V. Paatero. *Introduction to Complex Analysis*.
         Providence, RI: AMS Chelsea Publishing, 2007.

[P]   G. Polya. *Mathematics and Plausible Reasoning*, vol. 1. Prince-
         ton, NJ: Princeton University Press, 1990.

[Sp]   M. R. Spiegel. *Complex Variables*. Schaum's Outline Series.
         New York: McGraw-Hill, 1968.

[St]   J. Stewart. *Calculus: Concepts and Contexts*. Pacific Grove, CA:
         Brooks/Cole, 2001.

[Ta]   A. E. Taylor. A geometric theorem and its applications
         to biorthogonal systems. *Bull. Am. Math. Soc.* 53 (1947),
         pp. 614–616.

[TO]   T. F. Tokieda. Mechanical ideas in geometry. *Am. Math. Monthly*
         105 (8) (1998), pp. 697–703.

[To]   L. F. Toth. *Lagerungen in der Ebene auf der Kugel und im Raum*.
         Berlin: Springer-Verlag, 1953.

[U]   V. A. Uspenski. *Some Applications of Mechanics to Mathemat-
         ics*. New York: Pergamon Press, 1961.

INDEX